BOTANICAL AND HORTICULTURAL BOOKS:

A Price Guide

BOTANICAL AND HORTICULTURAL BOOKS:

A Price Guide

Compiled by
Carol Barnett

The Spoon River Press

First published March, 1991
Reprinted, June, 1995

The publisher has expended reasonable care in compiling the information found in this first edition of *Botanical and Horticultural Books: A Price Guide*; and therefore the publisher does not, nor do agents of the publisher, assume any liabilities of any kind whatsoever to any party for any damages assumed by such party to be a result of errors or omissions in this edition of *Botanical and Horticultural Books: A Price Guide*, regardless of the cause of such errors or omissions.

Published by
The Spoon River Press
2319-C West Rohmann Avenue
Peoria, Illinois 61604

ISBN 0-930358-09-0

Manufactured in the United States of America

Table of Contents

Introduction

This guide was compiled to aid in the evaluation and pricing of used and rare books in the fields of Botany and Horticulture. Its aim is to inform dealers, collectors, and librarians of current prices through a most reliable source – dealers' catalogues. Included are over 2,500 entries from the catalogues of booksellers in the United States, Great Britain, and Canada.

All entries are arranged alphabetically by author. The letter-by-letter style of alphabetizing has been used. In the case of anonymously published works, the title is entered in the general alphabetical list. To aid ease of reference, an attempt was made to standardize the entries. Many of the entries have been taken from the catalogues of book dealers in the United Kingdom. Thus, the prices for each entry are given in U.S. dollars and pounds sterling, the pound being converted at the exchange rate in effect at the time the catalogue was issued. Whenever possible, more than one dealer's price has been given for a book. However, an entry carrying only one price is not necessarily scarce or rare. Such titles have usually been noted.

Dates appearing in parentheses are on the verso of the title page of the work. Certain titles may be listed under the translator or an institution. Illustrations are not listed unless they constitute the body of a work.

Following the price of each book is a code in parentheses. This code designates the dealer from whose catalogue the listing was taken. The date of the catalogue's issue also is given. The dealer index at the back of this book keys these codes to specific dealers. The addresses of the dealers also are given. They may be contacted for further information or to request catalogues.

Condition is of the utmost importance in setting the price of any book – old or new. Items of condition and other items of book description were sometimes abbreviated in the entries cited in this reference. A Key to Abbreviations and Glossary of Terms Used has been included at the back of this book.

No books published after 1975 have been included in this reference. Some dealers may have multiple copies of newer books. Also, prices for many such books are somewhat low.

I wish to thank those dealers who so generously aided me in this endeavor.

A

"A Hermit." The Garden in the Wilderness. NY, 1909. 210pp. Decorated cover. VG. **$12.00\£7.00** *(RW 11/86)*

A Manual of American and European Plants. Funk & Wagnall's, 1900. 16mo. F in VG dj. **$10.00\£6.00** *(PBx 11/89)*

Abbot, Charles C. Travels in a Tree Top. Phil, 1898 (1894). T.e.g. 215pp. Slght spine fade else F. **$35.00\£19.00** *(Hrt 12/89)*

Abercrombie, J. The Gardener's Pocket Journal. London, 1807. 10th ed. 324pp. 12mo, hlf clf. G. **$54.00\£30.00** *(WW 1/90)*

Acton, Harold. Great Houses of Italy: The Tuscan Villas. NY, 1973. 388pp. Almost As New in dj. **$75.00\£42.00** *(Hrt 12/89)*

Adams, C.D. Flowering Plants of Jamaica. Mona, Jamaica, 1972. 848pp. Orange cloth. As New in dj. **$45.00\£25.00** *(BB 12/88)*

Adams, Fred. Conquest of the Tropics. G.C., Dblday, 1914. 368pp. Cloth light wear. **$20.00\£11.00** *(AB 2/89)*
Another copy. 368pp. Doubleday, 1914. VG. **$20.00\£11.00** *(GW 12/88)*

Adams, H.S. Flower Gardening. NY, 1913. 253pp. G. **$14.00\£8.00** *(BF 2/90)*

Adams, H.S. Making a Rock Garden. NY, 1912. 12mo. 52pp. 8 half tones. VG in dj. **$10.00\£6.00** *(RW 11/86)*

Addison, G. Malayan Orchid Hybrids: First Supplement. Singapore, 1961. 90pp. VG in dj. **$28.00\£16.00** *(GW 12/88)*

Adriance, G.W. & Brison, F.R. Propagation of Horticultural Plants. NY, 1939. [ix], 314pp. VG. **$20.00\£11.00** *(RS 2/89)*

Agardh, J.G. Species Genera et Ordines Algarum, Vol. 2: Florideae, Parts I & II Fasc. I. Lund, 1851-52. 535pp. 2 prts, 8vo. Orig wrpprs, neatly rebckd. **$43.00\£24.00** *(WW 1/90)*

Ahern, G. Compilation of Notes on the Most Important Timber Species of the Philippine Islands. Manilla, 1901. 112pp. Worn covers. **$35.00\£19.00** *(GW 8/89)*

Ahumada, Z. Revision de las Aristolochiaceae Argentinas. 1967. 151pp. Pbnd. G. **$12.00\£7.00** *(PBx 11/89)*

Aicardi, D. I Garofani Rifiorenti. San Remo, 1929. Brds. 275pp. G. **$18.00\£10.00** *(PBx 11/89)*

Aicardi, D.D. Le Rose: Moderne Cultivate ed Allevate. Rome, 1951. 12mo. 469pp. Sft brds. Handwritten ded. to Boerner. G. **$24.00\£13.00** *(PBx 11/89)*

Aiken, George. Pioneering with Fruits and Berries. NY, 1936. [xiv], 94pp. VG. **$15.00\£8.00** *(RW 6/89)*

Aiken, George. Pioneering with Wildflowers. Prentice, 1968. 208pp. VG in dj. **$10.00\£6.00** *(GW 12/88)*
Another copy. Putney, VT. Privately printed, 1933. 1st ed. [xvi], 124pp. VG. **$15.00\£8.00** *(JI 4/89)*
Another copy. NY, 1968. 208pp. VG in dj. **$15.00\£8.00** *(RW 6/89)*

Aiken, George. Pioneering with Wildflowers. Putney, 1935. 2nd prntng. 131pp. Faded brds else VG. **$18.00\£10.00** *(Hrt 12/89)*

(Aikin, John). The Woodland Companion. London, 1802. 1st ed. 92pp. Contemporary quarter calf and marbled boards. Generally somewhat worn especially head & tail of spine & hinges. **$153.00\£85.00** *(ABx 3/89)*

Alamanni, L. & G. Rucellai. La Coltivazione di Luigi Alamanni e le Api di Giovanni Rucellai. Bologna, 1746. [iv, xxxvi], 261pp. Orig brds, uncut. Nice cpy. **$225.00\£125.00** *(WW 1/90)*

Albaugh, Ben. The Gardenette. Cincinnati, 1920. 3rd ed. 138pp. Light wear. VG. **$18.00\£10.00** *(AB 2/89)*
Another copy. Cincinnati, 1917. 3rd ed. 138pp. VG. **$10.00\£6.00** *(RW 6/89)*

Albee, H.R. Hardy Plants for Cottage Gardens. NY, 1910. [vi], 309pp. lght wr. **$23.00\£13.00** *(RS 2/89)*

Aldrovandus, V. Illustrazione del Quarto Volume dell' Erbario di Ulisse Aldrovandi. Venice, 1911. 131pp. Wrpprs. G. **$36.00\£20.00** *(WW 1/90)*

Allan, Mea. E.A. Bowles and His Garden at Myd-Delton House 1865-1954. London, 1973. 264pp. F in dj. **$30.00\£17.00** *(BF 2/90)*

Allen, Betty. Some Common Trees of Malaya. Singapore, 1961. 2nd ed., wraps. 100pp. VG. **$14.00\£8.00** *(BB 12/88)*

Allen, H.H. New Zealand Trees and Shrubs. Auckland, 1928. [x], 188pp. Green cloth. VG. **$22.00\£12.00** *(BB 12/88)*

Allen, Lewis F. Rural Architecture. NY, 1860. Sm 8vo, brown cloth, 378pp. Worn with broken front joint. Severe shelf wear, split across spine. **$25.00\£14.00** *(MD 11/88)*

Allen, Mea. Tom's Weeds. London, 1970. 220pp. Good in dj. **$21.50\£12.00** *(ABx 3/89)*

Allen, Mea. William Robinson 1838-1935. London, 1982. 255pp. As New in dj. **$27.00\£15.00** *(ABx 3/89)*

Allen of Hurtwood, Lady and Susan Jellicoe. The Things We See: Gardens. London, 1953. 64pp. VG in wrpprs and dj. **$12.00\£7.00** *(BF 2/90)*

Allen, R.C. Roses for Every Garden. Rev. ed., 1956. G in dj. **$14.00\£8.00** *(PBx 11/89)*

Allwood, M.C. Carnations and All Dianthus. London, 1947. 3rd ed. **$27.00\£15.00** *(WW 1/90)*

Alston, A.H.G. The Ferns and Fern Allies of West Tropical Africa. London, 1959. 89pp. Ry 8vo, wrpprs. Used. **$27.00\£15.00** *(WW 1/90)*

Altschul, Siri von Reis. Drugs and Foods from Little Known Plants. Cambridge, 1975. 2nd prntng, 1973. [xii], 366pp. VG in dj. **$25.00\£14.00** *(BB 12/88)*

Amarican Iris Society - Alphabetical Iris Checklist. Baltimore, 1939. Ed by E.A.S. Peckham. [vii], 582pp. Lght wr. **$100.00\£56.00** *(RS 2/89)*

Ambasz, Emilio. Italy: The New Domestic Landscape. NY (MOMA), 1972. 430pp. VG in slightly discolored dj. **$35.00\£19.00** *(MD 2/90)*

American Box. The Old & the New in Flower Culture. Chicago, 1911. 24pp. Wraps, very light wear, a few pencil notes. **$20.00\£11.00** *(AB 4/89)*

American Rose Society. The American Rose Annual. Vols 6-15. 1921-1930. Clean, crisp set from prvt lib. **$85.00\£47.00** *(PBx 11/89)*

American-Moniger Greenhouse Mfg. Corp. American Greenhouses. Nd, 1920s. 384pp. Folio, pictorial brds, some staining. **$200.00\£111.00** *(Hrt 12/89)*
Another copy. Chicago, 1928. 384pp. Oversized folio. Fore edge stnd else VG. **$150.00\£83.00** *(BF 2/90)*

Anderson, A.W. Coming of the Flowers. NY, 1951. 267pp. VG in dj. **$18.00\£10.00** *(AB 4/89)*

Anderson, Alice Sloane (ed). Our Garden Heritage. NY, 1961. 622pp. F in dj. **$20.00\£11.00** *(BF 2/90)*

Anderson, E. Bertram. Gardening on Chalk and Limestone. London, 1965. Garden Book Club. 95pp. Good. **$12.50\£7.00** *(ABx 3/89)*

Anderson, Frederick. How to Grow Herbs for Gourmet Cooking. NY, 1967. 1st ed. 280pp. Ownr's inscrp else F in wrn, soiled dj. **$11.00\£6.00** *(CB 1/90)*

Anderson Galleries. (Auction Catalogue). The Botanical and Genealogical Library of the late Robert C. Wilkins together with Frank Forester Collection of Warren E. Freeman. NY, 1921. 365 items. Wraps. **$20.00\£11.00** *(RW 11/86)*

Andrews, H.N. Index of Generic Names of Fossil Plants, 1820-1950. 1955. 262pp. Pbnd. G. **$10.00\£6.00** *(PBx 11/89)*

Anley, G. Alpine House Culture for Amateurs. London, 1938. Trifle wrn, ink annotations. G. **$18.00\£10.00** *(WW 1/90)*

Anley, Gwendolyn. Alpine House Culture for Amateurs. London, 1938. 1st edition. 188pp. Inscribed by author. **$11.50\£6.00** *(BB 12/88)*

Ansell, Mary. The Happy Garden. London. Casell, 1912. [viii], 224pp. Decorated cover. VG. **$25.00\£14.00** *(RW 11/86)*

Another copy. NY, 1912. 224pp. Attractive decorated cloth, text underlined in spots. **$10.00\£6.00** *(AB 2/89)*

Arber, A. Goethe's Botany. 59pp. Chronica Botanica, 1946. Wraps. **$15.00\£8.00** *(GW 12/88)*

Arber, A. Monocotyledons. Cambridge, 1925. [xv], 258pp. Ry 8vo. Orig clth, spotted. **$45.00\£25.00** *(WW 1/90)*

Arber, A. The Natural Philosophy of Plant Form. Cambridge, 1950. [xiv], 247pp. 8vo. Trifle used. **$51.75\£28.00** *(WW 7/88)*

Arber, Agnes. Herbals. Cambridge U. Press, 1938. 2nd ed. [xxiv], 326pp. Covers slightly faded. G. **$117.00\£65.00** *(ABx 3/89)*

Arber, E.A.N. Plant Life in Alpine Switzerland. London, 1910. Trifle used. **$27.00\£15.00** *(WW 1/90)*

Architectural League of New York. 49th Annual Exhibition. NY, 1934. Spiral bnd wrps. VG. **$15.00\£8.00** *(MD 2/90)*

Ardéne, J.P.R. d'Traité des Renoncules. Paris, 1746. [x], 280pp. 1st ed. Lacks leaf of errata at end and 1st and last blanks. Sm piece mssng from inner corner of frontis else G. **$72.00\£40.00** *(WW 1/90)*

Another copy. 1746. Orig fll clf. Some worming to top outside mrgns of some pp else well preserved cpy. **$180.00\£100.00** *(DL 89/90)*

Armbrest, Duncan. The Beech Woods. Toronto, 1916. 1st ed. with Lady Eaton's bookplate. 120pp. 8vo. VG. **$25.00\£14.00** *(JI 12/88)*

Armitage, Ethel. A Country Garden. London, 1936. 1st ed. 226pp. G. **$45.00\£25.00** *(ABx 3/89)*

Armitage, Ethel. A Country Garden. NY, 1936. 1st Amer. ed. 226pp. Scttrd fxng, ownr's inscrp, bndng sunned. **$30.00\£17.00** *(BF 2/90)*

Armitage, Ethel. Garden and Hedgerow. London, 1939. 1st ed. 157pp. G in dj. **$45.00\£25.00** *(ABx 3/89)*

Arnold, F.H. Flora of Sussex. London, 1887. [xxiii], 118pp. Used, lacking back free endpaper. **$27.00\£15.00** *(WW 1/90)*

Arvidsson, Karl Axel. Millesgarden. Stockholm, 1953. English ed. 1/4 lthr on gilt dec brds. VG in dj. **$28.00\£16.00** *(Hrt 12/89)*

Asano, Kiichi & Gisei Takakuwa. Japanese Gardens Revisited. Rutland, VT, 1973. 165pp. F in slipcase. **$25.00\£14.00** *(BF 2/90)*

Ashberry, Anne. Miniature Gardens. London, 1951. 120pp. VG. **$25.00\£14.00** *(BF 2/90)*

Atchley, Shirley C. Wild Flowers of Attica. Oxford, 1938. [xix], [xxi], pp60. Good in dj. **$108.00\£60.00** *(ABx 3/89)*

Atkinson, G. Some Fungi from Alabama. Cornell, 1897. 50pp. Wraps. VG. **$10.00\£6.00** *(GW 12/88)*

Austin, A. The Garden That I Love. London, 1905. 2nd ed. [xiii], 146pp. Lght wr and soiling. **$30.00\£17.00** *(RS 2/89)*

Australian Rose Annual. 1928. Boards as issued, signed by Boerner. G. **$16.00\£9.00** *(PBx 11/89)*

Averill, Mary. Japanese Flower Arrangement (Ike-Bana) Applied to Western Needs. NY, 1913. 218pp. Corners slightly bumped. VG. **$50.00\£28.00** *(BF 2/90)*

Avery, G.S. & E.B. Johnson, et al. Hormones and Horticulture. NY & London, 1947. [xi], 326pp. Slght fxng. **$25.00\£14.00** *(RS 2/89)*

B

Backeberg, C. Die Cactaceae, Handbuch der Kakteenkunde, 1958-62. Vols. 1, 2, 4 & 6. Orig ed. VG. **$360.00\£200.00** *(DL 89/90)*

Bacon, Francis. Of Gardens. An Essay. London, John Lane, The Merrymount Press, 1902. 29pp + 15pp ads. **$12.50\£7.00** *(RW 11/86)*

Another copy. Boston, Merrymount, 1902. 29pp. Cntmprary bndng full crushes Morocco, by Bumpus. A.e.g, slght foxing end papers else F. **$95.00\£53.00** *(Hrt 12/89)*

Bagenal, N.B. Fruit Growing—Modern Cultural Methods. 1939. 1st ed. VG. **$23.00\£12.50 DL 89/90)**

Bailey, F. Manson. The Queensland Flora. Brisbane 1899-1902. 6prts w/index. Orig brds, clth back. Some slght wr to covrs else G set. Very scarce. **$270.00\£150.00** *(DL 89/90)*

Bailey, F. Manson. The Weeds and Suspected Poisonous Plants of Queensland. Brisbane, 1906. 245pp. Brds G. **$55.00\£30.00** *(WW 1/90)*

Bailey, L.H. (ed). Country Life in America, a Magazine for the Homemaker, the Vacation-seeker, the Gardener, the Farmer, the Nature-teacher, the Naturalist. Vol I, Nov. 1901-April 1902. Pblshd by Doubleday in 1902 w/ covering title page and index. Folio, cviii + 228pp. Title pg, index mildly wtrstnd, else near F. Individual cvrs and ads bnd in decor clth, somewhat rbbd and soiled. **$200.00\£111.00** *(BF 2/90)*

Bailey, L.H. & E.Z. Bailey. Hortus Second. NY, 1958. New ed. 778pp. Lght wr. **$50.00\£28.00** *(RS 2/89)*

Bailey, L.H. & Ethel Zoe Bailey. Hortus Second. NY, 1941. 778pp. Some scraping else VG. **$25.00\£14.00** *(CB 1/90)*

Bailey, L.H. Cyclopedia of American Horticulture. NY, 1900-1902. 4vols. [xxii], 2016pp. Orig rose patterned green clth. Gilt lettering (some wr at extrmities; some spine fdng; ownr's stmp on endpprs and foredges of pp; some hinge cracking). 1st ed. Scarce **$300.00\£167.00** *(RS 2/89)*

Bailey, L.H. Cyclopedia of American Horticulture. NY, 1906. 4th ed. 6 vols. [xiii], 2016pp. Spines and some edges sun faded; wrn as spine ends. **$150.00\£83.00** *(RS 2/89)*

Bailey, L.H. et al. How to Make a Flower Garden. NY, 1903. [xxii], 370pp. Lght wr and soiling. **$35.00\£19.00** *(RS 2/89)*

Bailey, L.H. Farm and Garden Rule-Book. NY, 1911. 1st ed. [xxiv], 587pp + 5pp ads. VG. **$15.00\£8.00** *(RW 6/89)*

Bailey, L.H. Field Notes on Apple Culture. NY, 1899. 90pp + 6pp ads. VG. **$20.00\£11.00** *(RW 6/89)*

Bailey, L.H. First Lessons with Plants. NY, 1911. 4th ed. 117pp. VG. **$15.00\£8.00** *(AB 2/89)*

Bailey, L.H. Garden Making. NY, 1913. [vii], 417pp. Lght soiling, wr at crnrs. **$25.00\£14.00** *(RS 2/89)*

Bailey, L.H. Garden-Making. NY, 1903. 417pp. VG. **$32.00\£18.00** *(BF 2/90)*

Bailey, L.H. Garden-Making. NY, 1913. 417pp. Light wear else VG. **$12.00\£7.00** *(AB 2/89)*
Another copy. NY, 1908. [vii], 417pp. In sl tender orig gilt blue cloth. VG. **$30.00\£17.00** *(Hrt 12/89)*

Bailey, L.H. Gardener's Handbook. Mcmillan, 1943. 292pp. Worn dj. **$12.50\£7.00** *(GW 8/89)*
Another copy. NY, 1934. 1st ed. 292pp. VG. **$15.00\£8.00** *(RW 6/89)*
Another copy. NY. 1943, (1934). 292pp. Bds tender. Internally VG. **$9.00\£5.00** *(Hrt 12/89)*

Bailey, L.H. Gardener's Handbook. NY, 1941. 292pp. Smll tr at top of spine. **$20.00\£11.00** *(RS 2/89)*

Bailey, L.H. Hortus, A Concise Dictionary of Gardening and General Horticulture. NY, 1930. 652pp. E.p.s brwning, ownr's inscrp. Slightly rbbd else VG. **$32.00\£18.00** *(BF 2/90)*

Bailey, L.H. Hortus. NY, 1930. 1st ed. 652pp. Lght wr. **$30.00\£17.00** *(RS 2/89)*

Bailey, L.H. How Plants Got Their Names. NY, 1933. 1st ed. [vi], 209pp. VG in dj. **$20.00\£11.00 (RW 6/89)**

Bailey, L.H. Manual of Cultivated Plants. NY, 1925. 851pp. Almost New. **$25.00\£14.00** *(AB 2/89)*

Bailey, L.H. Manual of Gardening. NY, 1932. 15th prntng. [xvi], 539pp. Brds VG. **$25.00\£14.00** *(Hrt 12/89)*
Another copy. NY, 1910. 541pp. Dec clth. VG. **$35.00\£19.00** *(BF 2/90)*

Bailey, L.H. Principles of Vegetable-Gardening. NY, 1907. 6th ed. 458pp. Spine worn. Marked on endpaper, contents VG. **$16.00\£9.00** *(AB 2/89)*

Bailey, L.H. Talks Afield About Plants. Boston, 1894. Later prntng (1885). 173pp. Blue cloth, ex-lib with no marks. Hinges cracked. Quite scarce. **$22.00\£12.00** *(AB 2/89)*
Another copy. Boston, 1885. 1st ed. 173pp + 4pp ads. VG. **$20.00\£11.00** *(RW 6/89)*

Bailey, L.H. The Apple-Tree. NY, 1922. 117pp. VG. **$12.50\£7.00** *(RW 6/89)*

Bailey, L.H. The Cultivated Evergreens. Macmillan, 1928. 434pp. Shelfwear else VG. **$20.00\£11.00** *(GW 12/88)*
Another copy. NY, 1925. [xvii], 434pp. VG. **$30.00\£17.00** *(RW 6/89)*

Bailey, L.H. The Cultivated Evergreens. NY, 1923. 1st. [xvii], 434pp. Sl spine fade else VG. **$50.00\£28.00** *(Hrt 12/89)*

Bailey, L.H. The Garden Lover. NY, 1928. 154pp. 12mo, ownr's inscrp else VG. **$28.00\£16.00** *(BF 2/90)*

Bailey, L.H. The Garden Lover. NY, 1928. 1st ed. 154pp. Green cloth, front hinge weak, scarce. Nearly new. **$20.00\£11.00** *(AB 2/89)*
Another copy. NY, 1928. 1st ed. 154pp. VG. **$10.00\£6.00** *(RW 6/89)*

Bailey, L.H. The Garden of Gourds. Boston, 1958. 134pp. Some lght stning else VG. **$20.00\£11.00** *(CB 1/90)*

Bailey, L.H. The Holy Earth. NY, 1946. [x], 117pp. Wraps. VG. **$5.00\£3.00** *(RW 11/86)*

Another copy. NY, 1915. 1st ed. [vi], 171pp. VG. **$15.00\£8.00** *(RW 6/89)*

Bailey, L.H. The Nursery Book. NY, 1896. 3rd ed. 12mo. [xi], 365pp. From the library of Beatrix Farrand Jones. Wr at extremities, spine faded, some spotting. **$35.00\£19.00** *(RS 2/89)*

Bailey, L.H. The Nursery-Book. NY, 1891. 1st ed. 304pp. G. **$22.00\£12.00** *(CB 1/90)*

Bailey, L.H. The Principles of Fruit Growing. NY, 1907. 10th ed. [xvii], 516pp. Extremities worn. VG. **$19.00\£11.00** *(CB 1/90)*

Bailey, L.H. The Pruning Manual. NY, 1944. [xiv], 407pp. Large 12mo. Hinge repaired else VG. **$15.00\£8.00** *(AT 3/89)*

Bailey, L.H. The Standard Cyclopedia of Horticulture. NY, 1917. 2nd ed. 639pp. Spine fade, set w/minimal rub, else very F. **$320.00\£178.00** *(Hrt 12/89)*

Bailey, L.H. The Standard Cyclopedia of Horticulture. NY, 1928. 3 vols. [xxiv], 3639pp. Wear at edges, djs have some wear and stning. **$150.00\£83.00** *(RS 2/89)*

Bailey, M. Robin Hood's Barn. NY, 1922. 301pp. Clth & paper brds; stns on frnt cvr, in dj. **$20.00\£11.00** *(RS 2/89)*

Bailey, M.E. Robin Hood's Barn. NY, 1922. 301pp. VG. **$6.00\£3.00** *(AB 2/89)*

Another copy. NY, 1922. 301pp. Near F, faded spine on dj. **$18.00\£10.00** *(Hrt 12/89)*

Bailey, Virginia L. and Harold E. Woody Plants of the Western National Parks. Notre Dame, 1949. 274pp. Blue cloth. VG. **$35.00\£19.00** *(BB 12/88)*

Baird, B.M. Roses for Southern Gardens. U of N.C. Press. 1948. In casette. Handwritten ded. to Boerner. F in shelfwrn casette. **$30.00\£17.00** *(PBx 11/89)*

Baker, R.T. & H.G. Smith. A Research on the Eucalypts. Sydney, 1920. 2nd ed. [xv], 470pp. Rather used, refixed in bndng. **$126.00\£70.00** *(WW 1/90)*

Baker, Richard St. Barbe. Trees, A Book of the Seasons. London, 1940. G. **$11.00\£6.00** *(ABx 3/89)*

Baker, Tarkinton. Yard and Garden. Indianapolis, 1913. [xviii], 418pp. Pictorial blue cloth. Large 12mo G. **$15.00\£8.00** *(AT 3/89)*

Baldwin, Henry. The Orchids of New England. NY, 1884. Scarce. VG. **$45.00\£25.00** *(RW 11/86)*

Baldwin, W.K.W. Plants of the Clay Belt of Northern Ontario & Quebec. 1958. 324pp. Pbnd. G. **$20.00\£11.00** *(PBx 11/89)*

Ball, C.R. Grain Sorghums: Immigrant Crops. Wash., DC. USDA, 1913. 17pp. **$5.00\£3.00** *(AB 4/89)*

Ball, G.J. Better Sweet Peas. Chicago, 1930. [5], 210pp. 2nd ed. VG. **$19.00\£11.00** *(RS 2/89)*

Ball, J. The Distribution of Plants on the South Side of the Alps. Linnean Soc., 1896. VG. **$27.00\£15.00** *(DL 89/90)*

Bandoni, R.J. & A.F. Szczawinski. Guide to Common Mushrooms of British Columbia. 1964. 179pp. Pbnd. G. **$9.00\£5.00** *(PBx 11/89)*

Barber, Peter and C.E. Lucas Phillips. The Trees Around Us. Chicago, 1975. 191pp. VG in brght dj, trn. **$25.00\£14.00** *(CB 1/90)*

Barclay, D., H.M. Bolus and E.J. Steer. A Book of South African Flowers.

Cape Town, 1925. [xvii], 174pp. Orig buckram. G. **$81.00\£45.00** *(WW 1/90)*

Barker, Frank. The Cream of Alpines. London, 1958. 86pp. G in dj. **$18.00\£10.00** *(ABx 3/89)*

Barnes, James & William Robinson. Asparagus Culture. NY, nd (early 1900s). 84pp. 12mo. Some bubbling to bck cvr, edge wrn, soiled. **$25.00\£14.00** *(CB 1/90)*

Barraclough, Daphne. A Flower-Lover's Miscellany. London, 1961. 160pp. F in rbbd dj. **$18.00\£10.00** *(BF 2/90)*

Barrell, Joseph. Flora of the Gunnison Basin. Gunnison, Saguache & Hinsdale Counties. Colorado. Rockford, 1969. 494pp. Green cloth. VG. **$33.00\£18.00** *(BB 12/88)*

Barron, Leonard. Lawns and How to Make Them, Together with the Proper Keeping of Putting Greens. NY, 1909. 174pp. Mild fxng, generally VG. **$20.00\£11.00** *(BF 2/90)*

Barry, P. The Fruit Garden. NY. Moore, 1859. VG. **$10.00\£6.00** *(RW 11/86)*

Barton, J. A Lecture on the Geography of Plants. London, 1827. [4], 94pp. Orig brds, neatly rebacked, uncut. **$54.00\£30.00** *(WW 1/90)*

Barton, J.G. Wildflowers. London, 1963. 4to. VG in dj. **$10.00\£6.00** *(RW 11/86)*

Barton, L.V. & W. Crocker. Twenty Years of Seed Research. London, 1948. Faded. **$22.00\£12.00** *(WW 1/90)*

Bartram, William. Travels through North & South Carolina, Georgia, East & West Florida. London, 1792. [xxiv], 520pp. Early half leather on marbled boards, corners worn. Some brwning of pp else nice, tight copy. **$1400.00\£777.00** *(Hrt 12/89)*

Bartrum, Douglas. Evergreens for Your Garden. London, 1967. Garden Book Club. 187pp. G in dj. **$11.00\£6.00** *(ABx 3/89)*

Bary, A. de. Comparative Anatomy of the Vegetative Organs of the Phanerogams and Ferns. Oxford, 1884. [xvi], 659pp. Ry 8vo, orig hlf morocco. Spine rather worn. **$54.00\£30.00** *(WW 1/90)*

Bastin, Harold. Plants without Flowers. London, 1955. 1st ed. [xii], 148pp. Sl cover fading else VG. **$15.00\£8.00** *(JI 4/89)*

Bateman, J. The Orchidaceae of Mexico and Guatemala. London, 1974. [viii], 12, 4, 40pp. Imp folio, clth. Slightly reduced repro. G. **$315.00\£175.00** *(WW 1/90)*

Bates, Alfred. The Gardener's Third Year. NY, 1941. [ix], 310pp. F in dj. **$24.00\£13.00** *(Hrt 12/89)*

Bates, H.E. A Love of Flowers. London, 1971. 160pp. G. **$25.00\£14.00** *(BF 2/90)*

Batson, Mrs. Stephen. The Summer Garden of Pleasure. Chicago, 1909. 231pp. Spine sunned, else VG. Bookplate removed. **$25.00\£14.00** *(AB 4/89)*

Batson, Wade T. Wild Flowers in South Carolina. Columbia. U of S Carolina Press, 1964. 1st ed. Signed by the Author. [vi], 146pp. Covers slightly worn. **$25.00\£14.00** *(JI 7/88)*

Battiscombe, E. Trees and Shrubs of Kenya Colony. Nairobi, 1936. Rev and enlrgd ed. [xii], 204pp. Boards. Sl sunning else VG. **$25.00\£14.00** *(JI 12/88)*

Battles, Marjorie & C. Dickey. Fifty Blooming Years 1913-1963. Np, 1963. 98pp. Paper over boards. G only. **$19.00\£11.00** *(CB 12/89)*

Battles, Marjorie G. & Catherine C. Dickey. Fifty Blooming Years, 1913-1963. A History of the Garden Club of America. NY, 1963. Ed. of 3000. F. **$22.00\£12.00** *(BF 2/90)*

Another copy. Np, 1963. 98pp. Ltd to 3000 cpies. G. **$19.00\£11.00** *(CB 1/90)*

Bawden, H.E. & J. Making a Shrub Garden. London, 1966. **$21.00\£12.00** *(WW 1/90)*

Beach, S.A., et al. The Apples of New York. Albany, 1905. 2 vols. 409, 360pp. F set. **$150.00\£83.00** *(WH 1/90)*

Beadle, D.W. Canadian Fruit, Flower & Kitchen Gardener. Canada, 1872. Recased, orig. spine. Scarce. Immaculate. **$85.00\£47.00** *(PBx 11/89)*

Bealby, J.T. Fruit Ranching in British Columbia. London, 1901. 1st ed. 196pp. Spine faded else G. **$81.00\£45.00** *(ABx 3/89)*

Beale, James H. The Evergreens. NY, 1960. 285pp. F. **$7.50\£4.00** *(WH 1/90)*

Bean, W.J. The Royal Botanic Gardens Kew. London, 1908. 2nd imp. 222pp. Light spotting on end papers else a very nice copy. **$81.00\£45.00** *(ABx 3/89)*

Bean, W.J. Trees and Shrubs Hardy in the British Isles. London, 1919-1933. 2nd ed. 3 vols. 3rd suppl. vol is 1st ed. Pencil annotations. Trifle used. **$45.00\£25.00** *(WW 1/90)*
Another copy. London, 1925. 4th ed. 2 vols. [xvi], 1424 pp total. Green clth. Slght bubbling to clth else VG. **$65.00\£31.00** *(CB 1/90)*

Bean, W.J. Trees and Shrubs Hardy in the British Isles. 1916 & 1933. 3 vols. Vols I & II: 1916, 2nd ed. Vol III: 1933, 1st ed., updating the originals. 688, 736, 517pp. Serviceable set, G. **$79.00\£44.00** *(ABx 3/89)*

Beard, P. Adventures in Dish Gardening. NY, 1930. 12mo. [xii], (2), 204pp. Wr at edges; edges drknd. **$12.50\£7.00** *(RS 2/89)*

Beattie, W. Celery Culture. NY, 1920. 143pp. Nearly New. **$16.00\£9.00** *(AB 2/89)*

Beck, Christabel. Fritillaries. London, 1953. 96pp. Lg 8vo. Scarce. G. **$90.00\£50.00** *(Hrt 12/89)*

Beddome, R.H. The Ferns of Southern India. Madras, 1873. 2nd ed. [xv], 88, [xv]pp. 4to. Bckrm. Rare orig prntng. Bndng rather used and mrgns of 2 leaves defect. **$198.00\£110.00** *(WW 1/90)*

Beebe, William. Edge of the Jungle. NY, 1921. 1st ed. 303pp. VG in aging dj. **$36.00\£20.00** *(Hrt 12/89)*

Beebe, William. The Book of Naturalists. NY, 1948. [xiv], 499pp. Bkplte & sig else VG. **$40.00\£22.00** *(Hrt 12/89)*

Beeton. The Book of Garden Management. Nd, c1895. 844pp. G. **$54.00\£30.00** *(ABx 3/89)*

Behme, Robert L. Bonsai, Saikei and Bonkei. NY, 1969. VG in dj. **$30.00\£17.00** *(Hrt 12/89)*

Behr, H.H. Synopsis of the Genera of Vascular Plants in the Vicinity of San Francisco. SF, 1884. 165pp. 12mo. Brick colored cloth, blind stamped. Neat ex-library. **$54.00\£30.00** *(BB 12/88)*

Bell, Robert. The Forests of Canada. Montreal, 1886. 15pp. Wraps. VG. **$20.00\£11.00** *(Hrt 12/89)*

Bellamy, D. Ethic Amusements. By Mr. Bellamy, rev by his son D. Bellamy. London, 1768. 4to. 260pp. Title pg followed by 41 plts, all rev images of Perrault's "Labyrinte", all hors-texte, (loose), blank, frontis, title, blank, (2) not numbrd that are prt of dedication to the King, ix, (1) direction for cuts, a title, blank. Possible pres cpy. Blue morocco glt dble bordering/outer, vase w/flower, bordered by 2 columns; inner, globe w/flwrs & acorns, bordered by single, dble tiered vases, red Moroccan label; diapers, floral design on spine. Rare paper ed w/fleur-de-lys watermrk. Slght shelfwr, slght offset to title. Else near F, a.e.g. contemp bkplte of Lord Camden. **$1850.00\£1027.00** *(Hrt 12/89)*

Benhall, A.P. The Trees of Calcutta. Calcutta, 1946. [cii], 513pp. G. **$36.00\£20.00** *(WW 1/90)*

Bennett, A.W. The Flora of the Alps: Being a Description of All. . . Indigenous to Switzerland; & of the Alpine Species. . . of France, Italy & Austria, including the Pyrenees. 1896. T.e.g., orig clth. Unusually well preserved, tght set. **$135.00\£75.00** *(DL 89/90)*

Bennett, Ida D. The Flower Garden. NY, 1905. 2nd prntng. 282pp. Ownr's name else VG. **$18.00\£10.00** *(BF 2/90)*

Bennett, Ida D. The Flower Garden. NY, 1909. [ix], 282pp. VG. **$10.00\£6.00** *(RW 6/89)*

Benrath, H. Die Welt der Rose. 1954. Lge 4to. 2 diff. dj (sic) laid in letter to Eugene Boerner from W. Kordes. G in fair dj. **$40.00\£22.00** *(PBx 11/89)*

Benson, Albert E. History of the Massachusetts Horticultural Society. 1929. Large 8vo. VG. **$15.00\£8.00** *(RW 11/86)*

Benson, Gilbert T. The Trees and Shrubs of Western Oregon. Stanford U., 1930. Contr. from the Dudley Herbarium – vol. 2, 170pp. Color map. Dj chipped at head of spine & front cover. Small tear first 12pp foredge. **$18.00\£10.00** *(BB 12/88)*

Bentham, G. Flora Hongkongensis. London, 1861. [20], li, 482pp. New cloth. **$252.00\£140.00** *(WW 1/90)*

Bergen, H. von. Versuch einer Monographie der China. Hamburg, 1826. [xii], 348pp. Wrpprs. W/o atlas. Wrn and trifle dustsoiled, stmp on title pge. **$45.00\£25.00** *(WW 1/90)*

Berkeley, Edmund and Dorothy. John Clayton. Chapel Hill, 1963. 1st ed. [xii], 236pp. VG. **$30.00\£17.00** *(JI 12/88)*

Berkeley, M.J. Introduction to Cryptogamic Botany. London, 1857. [ix], 604pp. Clth, trifle used and faded. **$54.00\£30.00** *(WW 1/90)*

Berkenhout, J. Clavis Anglica Linguae Botanicae. London, 1764. [xii], 215pp. Contemp. sheep. Library bkplte but no stmps. Good, clean copy. Scarce. **$63.00\£35.00** *(WW 1/90)*

Berrall, Julia S. The Garden. NY, 1966. 1st ed. 388pp. VG in somewhat wrn dj. **$35.00\£19.00** *(CB 1/90)*

Berrisford, Judith. Rhododendrons & Azaleas. London, 1973. New revised ed. 289pp. G in dj. **$27.00\£15.00** *(ABx 3/89)*

Berry, James B. Southern Woodland Trees. Chicago, 1924. [vii], 214pp. Card pocket removed, no other lib. marks. Owner's blind stamp on several pages. **$12.00\£7.00** *(BB 12/88)*

Berry, James B. Western Forest Trees. Chicago, 1926. [x], 212pp. 96 photos and drawings. VG. **$14.00\£8.00** *(BB 12/88)*

Bessey, Charles E. Grasses and Forage Plants of Nebraska. Lincoln, 1888. 15pp. Orig green wrappers. F. **$17.50\£10.00** *(WH 1/90)*

Best, K.F. and A.C. Budd. Common Weeds of the Canadian Prairies. Ottawa. CDA, 1963. 72pp. Wraps. VG. **$15.00\£8.00** *(JI 4/89)*

Beston, Henry. Northern Farm. NY, 1948. 246pp. Light wear else VG. **$9.00\£5.00** *(AB 2/89)*

Betts, Edwin M. & Hazelhurst B. Perkins. Thomas Jefferson's Flower Garden at Monticello. Richmond, 1941. 56pp. Some fxng frnt/rear. **$30.00\£17.00** *(Hrt 12/89)*

Betts, Edwin M. & Hazelhurst B. Perkins. Thomas Jefferson's Flower Garden at Monticello. Richmond, 1941. 56pp. VG. **$20.00\£11.00** *(RW 6/89)*

Bianchini, F. and F. Corbetta. The Complete Book of Fruits and Vegetables. NY, 1975. 1st US ed. 304pp. As New in dj. **$30.00\£17.00** *(JI 4/89)*

Bichard, John & David McClintock. Wild Flowers of the Channel Islands. London, 1975. 1st ed. [xii], 68pp. VG in dj. **$15.00\£8.00** *(JI 4/89)*

Bicknell, C. Flora of Bordighera and San Remo. Bordighera, 1896. [viii], 345pp. Lacks map, shows signs of use. **$27.00\£15.00** *(WW 1/90)*

Biddle, D. & D. Blom. Garden Gossip Chronicles of Sycamore Valley. GC, 1936. 1st. 95pp. Cloth light wear. **$15.00\£8.00** *(AB 2/89)*

Biddle, Moncure. A Christmas Letter—Some Flower Books and Their Makers. Phil, 1945. 49pp. VG in slightly wrn wrpprs. **$12.00\£7.00** *(BF 2/90)*

Billings, W.D. Nevada Trees. Reno, 1954. 2nd ed. 125pp. Photo wraps. Key. Library stamps. **$14.00\£8.00** *(BB 12/88)*

Bingham, M. Flora of Oakland County, Michigan. 155pp. Cranbrook, 1945. Wrs. VG. **$15.00\£8.00** *(GW 8/89)*

Bingham, Marjorie T. The Flora of Oakland County, Michigan. Cranbrook Inst., 1945. Bulletin #22. 155pp. Orig clth. Nr F. **$18.00\£10.00** *(Hrt 12/89)*

Birdsey, M.R. The Cultivated Aroids. Berkeley, 1951. (3), 140pp. Sgnd by author. Some wr at crnrs, lght spine fading. **$40.00\£22.00** *(RS 2/89)*

Birdseye, C. & E. Growing Woodland Plants. NY. Oxford UP, 1951. 223pp. F in dj. **$12.50\£7.00** *(WH 1/90)*

Bisset, P. The Book of Water Gardening. NY, 1924. 2nd ed. 205pp. Gilt and blue decor clth; a few smudge mrks on pp; dmp stning in lower rght crnr of back pp. **$35.00\£19.00** *(RS 2/89)*

Bisset, Peter. The Book of Water Gardening. NY, 1924. 2nd ed., rev. 205pp. Dec cover. VG. **$30.00\£17.00** *(RW 6/89)*

Biswas, K. Plants of the Darjeeling & the Sikkim Himalayas. Bombay, 1966. Vol I. 540pp. VG. **$55.00\£31.00** *(GW 8/89)*

Black, J.M. The Naturalised Flora of South Australia. Adelaide, 1909. 1st ed. 192pp. 12mo. Green cloth. Scarce. **$42.00\£23.00** *(BB 12/88)*

Blaikie, Thomas. Diary of a Scotch Gardener at the French Court at the End of the Eighteenth Century. NY, 1932. 1st Amer ed. 256pp. Ex-garden club, few internal stns and mrks, slightly flaked clth. **$30.00\£17.00** *(BF 2/90)*

Blair, Laurence. The Garden Clinic. NY, 1940. 146pp. VG in slightly soiled dj. **$20.00\£11.00** *(BF 2/90)*

Blake, Peter. The House in the Museum Garden. NY, 1949. 16pp. Pictorial wraps. 8vo. VG. **$12.50\£7.00** *(AT 3/89)*

Blake, S. A Revision of the Genus Viguiera (Asteraceae). Harvard, 1918. 203pp. Wraps. Front cover torn. **$16.00\£9.00** *(GW 12/88)*

Blake, S.F. Geographical Guide to Floras of the World. Wash., DC. USDA, 1971. Part II only. 742pp. VG. **$18.00\£10.00** *(AB 4/89)*

Another copy. USDA, 1961. 2 vols. 1078pp. Orig ed. VG. **$36.00\£20.00** *(GW 12/88)*

Blakeslee, A.F. The Genus Datura. NY, 1959. 289pp. As New. **$35.00\£19.00** *(AB 4/89)*

Blanchan, Neltje. The American Flower Garden. London, 1909. 368pp. Slightly faded in parts, small nick in rear hinge. G. **$99.00\£55.00** *(ABx 3/89)*

Blanchan, Neltje. The American Flower Garden. NY, 1909. 368pp. Rear cvr heavily stned. **$30.00\£17.00** *(BF 2/90)*

Blasdale, Walter. The Cultivated Species of Primula. Berkeley & LA, 1948. [xii], 284pp. 8vo. Green cloth. Light spotting lower right corner else VG. Scarce. **$45.00\£25.00** *(CB 12/89)*

Blatter, E. & W.S. Millard. Some Beautiful Indian Trees. London, 1937. [x], 110pp. G. **$45.00\£25.00** *(WW 1/90)*
Another copy. 1937. 1st ed. VG. **$54.00\£30.00** *(DL 89/90)*

Blatter, E. Beautiful Flowers of Kashmir. London, 1928-29. 2 vols. Vol 2 is a latter issue in a non-uniform faded binding. G. **$117.00\£65.00** *(WW 1/90)*

Blomfield, R. & Thomas F. Inigo. The Formal Garden in England, 1892. 1st ed. Orig bckrm. VG. **$63.00\£35.00** *(DL 89/90)*

Blomfield, Reginald & F. Inigo Thomas. The Formal Garden in England. London, 1892. 1st ed. [xii], 244pp. Light foxing throughout. Label on spine torn. G. **$108.00\£60.00** *(ABx 3/89)*

Bloom, Adrian. Conifers for Your Garden. London, 1972. 145pp. G. **$11.00\£6.00** *(ABx 3/89)*

Blunt, Wilfred. The Compleat Naturalist. NY, 1971. 1st Amer. 256pp. As New. **$45.00\£25.00** *(Hrt 12/89)*

Blunt, Wilfrid. The Art of Botanical Illustration. London, 1950. 1st ed. 304pp. G in dj. **$99.00\£55.00** *(ABx 3/89)*
Another copy. London, 1950. 1st ed. 304pp. VG. **$80.00\£44.00** *(RW 6/89)*

Blunt, Wilfrid. The Compleat Naturalist. London, 1971. 256pp. G in dj. **$43.00\£24.00** *(ABx 3/89)*

Blunt, Wilfrid. Tulipomania. London, 1950. 1st ed. 32pp. King Penguin #44. G in dj. **$27.00\£15.00** *(ABx 3/89)*
Another copy. Harmondsworth, 1950. 1st ed. King Penguin #44. 32pp. VG in torn dj. **$30.00\£17.00** *(CB 12/89)*

Boardman, Samuel Lane. The Naturalist of the Saint Croix. Bangor. Priv prntd, 1903. Ltd to 500 cpies. [xi], blank, (1) contents, blank (1) list of plts, blank, 351pp, (1) colophon. Hlf lthr on reddish mrbld paper. Rare. F. **$340.00\£189.00** *(Hrt 12/89)*

Boas, I.H. The Commercial Timbers of Australia. Melbourne, 1947. 1st ed. [viii], 344pp. VG. **$25.00\£14.00** *(JI 12/88)*

Boatfield, Jeffrey M. Soft Fruit. London, 1961. 207pp. VG. **$14.00\£8.00** *(Hrt 12/89)*

Boggs, Kate Boggett. Prints and Plants of Old Gardens. Richmond, 1932. 101pp. Folio. Dec clth, spine ends wrn else VG. **$60.00\£33.00** *(BF 2/90)*

Boggs, Kate Doggett. Prints and Plants of Old Gardens. Richmond, 1932. Thin 4to, decorative rust and green cloth. 101pp. Small tear to edge of rear cover & a few page edges else VG. **$45.00\£25.00** *(MD 7/89)*

Bois, D. Les Plantes Alimentaires. Paris, 1927-37. 4 vols. Vol. I jnts rather loose. G. **$126.00\£70.00** *(WW 1/90)*

Bois, E. & A.M. Trechslin. Roses. 1962. 4to. F. **$45.00\£25.00** *(PBx 11/89)*

Bonnier, Gaston. Name This Flower. NY, 1917. 331pp. Cloth lt wear. **$12.00\£7.00** *(AB 2/89)*

Boothman, Stuart. The Alpine House and Its Plants. Bedford, England, 1938. 131pp. Scarce. Foxing on edges, end papers browning. G. **$18.00\£10.00** *(BB 12/88)*

Bor, Eleanor. Adventures of a Botanist's Wife. London, 1952. 204pp. G. **$21.50\£12.00** *(ABx 3/89)*

Bor, N.L. Some Beautiful Indian Climbers and Shrubs. Bombay, 1954. [viii], 286pp. G. **$45.00\£25.00** *(WW 1/90)*

Bor, N.L. The Grasses of Burma, Ceylon, India and Pakistan (excluding Bambuseae). 1960. Orig clth. VG. **$63.00\£35.00** *(DL 89/90)*

Borg, J. Cacti. London, 1963. 512pp. 3rd ed. E.p.s. tape-marked else G. **$27.00\£15.00** *(WW 1/90)*
Another copy. London, 1956. 2nd ed. 487pp. Green clth. G. **$25.00\£14.00** *(CB 1/90)*

Borgesen, F. Marine Algae from the Canary Islands. Copenhagen, 1925-36. 631pp. G. **$72.00\£40.00** *(WW 1/90)*

Boston Society of Landscape Architects, Year Book for 1929. 4to. X-lib o/w VG. **$45.00\£25.00** *(RW 11/86)*

Botke, J. and C.M. v.d. Slikke. Lelies en Leliecultuur. 's-Gravenhage, 1942. 155pp. Glt decor clth, spine faded; some overall wr. **$45.00\£25.00** *(RS 2/89)*

Boulger, Prf. G.S. & Mrs. Henry Perrin. British Flowering Plants. London, 1914. Limited ed. of 1000 copies. 4 vols. White buckram spines slightly darkened. G. **$288.00\£160.00** *(ABx 3/89)*

Bowen, Louise deKoven. Baymeath. Np, 1944. 137pp. Inscribed by author. Spine worn. **$20.00\£11.00** *(RW 6/89)*

Bower, F.O. Primitive Land Plants. NY, 1959. [xiv], 658pp. Ry 8vo, buckram, trifle used. **$29.00\£16.00** *(WW 1/90)*

Bower, F.O. Size and Form in Plants. London, 1930. [xi], 232pp. G. **$36.00\£20.00** *(WW 1/90)*

Bowers, C. Rhododendrons & Azaleas: Their Origins, Cultivation & Development. McMillan, 1936. 549pp. **$45.00\£25.00** *(GW 12/88)*

Bowers, C. The Development of Pollen and Viscid Strands in Rhododendron Catawbiense. 1931. 28pp. Wraps. VG. **$8.50\£5.00** *(GW 12/88)*

Bowles, E.A. A Handbook of Crocus and Colchicum for Gardeners. London, 1952. Revised ed. 222pp. VG in dj. **$39.50\£22.00** *(ABx 3/89)*

Bowles, E.A. A Handbook of Crocus and Colchicum for Gardens. London, 1924. 1st ed. [xii], 185pp. 8vo. Lght wr; smll spot of cvr, ownr's stmp on pp foredges. **$47.00\£26.00** *(RS 2/89)*

Bowles, E.A. A Handbook of Narcissus. London, 1934. 248pp. G. **$27.00\£15.00** *(WW 1/90)*

Bowles, E.A. My Garden in Spring. NY, 1920. 307pp. Light wear. **$22.00\£12.00** *(AB 2/89)*
Another copy. NY, 1914. Decr clth, t.e.g., excellent cond. **$40.00\£22.00** *(PBx 11/89)*

Bowles, E.A. My Garden in Spring. Pawlet, 1971. [xx], 308pp. **$18.50\£11.00** *(RS 2/89)*

Bowles, E.A. My Garden in Spring. Pawlet, VT, 1971. Facs reprnt. 308pp. F in clear plastic dj. **$20.00\£11.00** *(BF 2/90)*

Bowles, John. Versailles Illustrated. London, nd, 1720s. Prnted & sold by John Bowles & Son, obling 4to, title pp, with 4 views, coat of arms & decor frame for prntr's inscrip. 22 plts of various views, (12 single image, 11 dble, 1 three) last 7 plts are the Labyrinth; plan, description, entrance & 39 fables. All prose, verse have been translated and inscribed on engrvngs, top & bttm. But for very lght offset to a number of plts, a VG cpy in contemp mrbld brds. **$1850.00\£1028.00** *(Hrt 12/89)*

Boyd, Lizzie Edmunds. England, Her People and Her Gardens. Boston, 1940. 31pp. Very nice copy. **$20.00\£11.00** *(Hrt 12/89)*

(Boyle, Eleanor Vere). "E.V.B.", A Garden of Pleasure. London, 1895. 3rd ed. 220pp. Slightly loose, else G. **$25.00\£14.00** *(ABx 3/89)*

(Boyle, Eleanor Vere). "E.V.B.", Seven Gardens and a Palace. London, 1900. 298pp. VG. **$20.00\£11.00** *(RW 6/89)*
Another copy. London, 1900. 298pp. End pps frnt/bck missing. Rebnd in very tender, stained full imitation leather else VG. **$12.00\£7.00** *(Hrt 12/89)*

(Boyle, Eleanor Vere). "E.V.B.", Sylvana's Letters to an Unknown Friend. London, 1900. 190pp. G. **$36.00\£20.00** *(ABx 3/89)*
Another copy. London, 1900. 190pp. T.e.g., tender & worn brds. **$12.00\£7.00** *(Hrt 12/89)*

Boyle, L.M. Out West Growing Cymbidium Orchids. N.p., 1952. 526pp. **$35.00\£19.00** *(RS 2/89)*

Bradley, Edith & Bertha La Mothe. The Lighter Branches of Agriculture. The Woman's Library. Vol VI. NY, nd (c1900). [xx], 346pp. Decorated cover. VG. **$12.50\£7.00** *(RW 11/86)*

Bradley, R. Le Calendrier des Jardiniers. Paris, 1750. Marbled endpapers, plates need refolding else VG in contemporary leather. Gilt decrted spine. **$320.00\£178.00** *(Hrt 12/89)*

Braithwaite, R. The Sphagnaceae. London, 1880. [8], 91pp. Orig clth, trifle wrn. Inscription and stamp on ep. F.f.e.p. cut away. Occ minor dustsoiling, but generally G. Scarce colored issue. **$135.00\£75.00** *(WW 1/90)*

Bramwell, David and Zoe. Wild Flowers of the Canary Islands. London, 1974. 1st ed. [x], 262pp. A few nicks else F in dj. **$25.00\£14.00** *(JI 7/88)*

Brandis, D. Indian Trees—An Account of Trees, Shrubs, Woody Climbers, Bamboos and Palms Indigenous or Commonly Cultivated in the British Indian Empire. 1906. 1st ed. Scarce. VG. **$72.00\£40.00** *(DL 89/90)*
Another copy. Dehra Dun, 1971. G. **$36.00\£20.00** *(WW 1/90)*

Brandis, D. The Forest Flora of Northwest & Central India, Commenced by the Late J. Lindsay Stewart. 1874. 608pp. Hlf morocco, t.e.g. Very scarce. Atlas vol not present. G. **$55.00\£31.00** *(PBx 11/89)*
Another copy. London, 1874. [xxi], 608pp. New brds (libry stmp). G. **$54.00\£30.00** *(WW 1/90)*

Braun, E. Lucy. Deciduous Forests of Eastern North America. Phila, 1950. 596pp. F. **$37.50\£21.00** *(WH 1/90)*

Braun, E. The Monocotyledoneae. OSU, 1967. 464pp. VG in dj. **$20.00\£11.00** *(GW 8/89)*

Braybrooke, Neville. London Green. London, 1959. 200pp. G in torn dj. **$21.50\£12.00** *(ABx 3/89)*

Breck, J. The Flower Garden. Boston, 1851. 1st ed. 336pp. Orig clth, wrn at crnrs; spine ends chppd away; spine splitting, reglued. Occ fxng. **$65.00\£36.00** *(RS 2/89)*

Breedlove, D. The Systematics of Fuchsia Section Encliandra (Onagraceae). CA, 1969. 69pp. Wrs. **$8.50\£5.00** *(GW 8/89)*

Brewer, James. A New Flora of the Neighbourhood of Reigate. London, 1856. [viii], 194pp. Top of spine worn else G. Quite rare. **$81.00\£45.00** *(ABx 3/89)*

Brewster, Kate L. The Little Garden for Little Money. Boston, 1924. 109pp. 12mo. VG. **$10.00\£6.00** *(RW 6/89)*

Bridgeman, Thomas. Flower Gardening. Phil, nd, early 1900s. 166pp. Head/foot of spine worn else VG. **$17.00\£9.00** *(CB 1/90)*

Bridgeman, Thomas. Young Gardener's Assistant. NY, 1857. 515pp. New edition with appendix. Extremities worn else VG. **$50.00\£28.00** *(AB 4/89)*
Another copy. NY, 1865. New edition. 189pp. VG. **$45.00\£25.00** *(RW 6/89)*

Briggs, Loutrel W. Charleston Gardens. Columbia. U of S.C. Press, 1951. 156pp. F. **$22.00\£12.00** *(BF 2/90)*

Brillat-Savarin, J.A. The Physiology of Taste. NY, 1926. 361pp. F. **$20.00\£11.00** *(PBx 11/89)*

Brilmayer, B. All About Vines & Hanging Plants. 1962. 384pp. **$16.00\£9.00** *(PBx 11/89)*

Brimble, L.J.F. Flowers in Britain. London, 1952. [x], 393pp. VG in dj. **$10.00\£6.00** *(RW 6/89)*

Brimble, L.J.F. Flowers in Britian. London, 1945. [x], 393pp. Inscrp on e.p., in wrn dj. **$18.50\£11.00** *(RS 2/89)*

Briscoe, Rufus J. The Two Oldest Trees: One Dead, One Living. Riverside, CA, 1914. 63pp. 12mo. Cloth backed thick boards. F.f.e.p. missing. **$17.50\£10.00** *(BB 12/88)*

British Bryological Society. Transactions, Vols. 1-4. London, 1947-65. Rather crude amateur bndng. **$90.00\£50.00** *(WW 1/90)*

Britten, J. European Ferns. Nd. 1st ed. 4to. 1/2 green clf. G. **$90.00\£50.00** *(DL 89/90)*

Britton, A. & N. Brown. An Illustrated Flora of the Northern U.S., Canada & The British Posessions. Scribner, 1913. 3 vols. 2052pp. Worn bindings. **$45.00\£25.00** *(GW 8/89)*

Britton, N. & J. Rose. Genus Epiphyllum and Its Allies. Wash. GPO, USNH, 1913. 12pp. Wraps, VG. **$8.00\£4.00** *(AB 4/89)*

Britton, N.L. & J.N. Rose. The Cactaceae. NY, 1963. Reprint of the 2nd ed. 4 vols bound as 2. Almost As New in djs. **$108.00\£60.00** *(ABx 3/89)*

Britton, N.L. & J.N. Rose. The Cactaceae. WA, 1919-23. Reprnt 1964. 1068pp. 4 vols in 2. G. **$93.00\£52.00** *(WW 1/90)*
Another copy. Wash, 1919-1923. 1st ed. 4 vols. Vol I: 1919, [vii], 236pp. Vol II: 1920, [vii], 241pp. Vol III: 1922, [vi], 258pp. Vol IV: 1923, [vii], 318pp. Becoming quite scarce. VG in orig brds with some staining. **$575.00\£319.00** *(Hrt 12/89)*

Britton, N.L. Catalogue of Plants Found in New Jersey. Trenton, 1889. 616pp. Wrpprs, G. **$27.00\£15.00** *(WW 1/90)*

Britton, N.L. et al. List of Pteridophyta & Spermatophyta Growing with TT Cultivation. Torrey Botanical Club, 1894. 377pp. Covers detached, well worn. Contents all present. **$20.00\£11.00** *(AB 4/89)*

Brockman, C. Frank. Flora of Mount Rainier National Park. Washington, 1947. 170pp. Wrappers. VG. **$10.00\£6.00** *(WH 1/90)*

Bromfield, Louis. Malabar Farm. NY, 1948. 405pp. A bit shaken, dj worn. **$5.00\£3.00** *(AB 2/89)*

Bromfield, Louis. Out of the Earth. NY, 1950. 305pp. As New in dj. **$7.00\£4.00** *(AB 2/89)*

Brookes, John. Room Outside. 1969. 192pp. Probable 1st ed of 1st book. VG in dj. **$35.00\£19.00** *(Hrt 12/89)*
Another copy. NY, 1970. 192pp. 2nd prntng. VG in bright, edge-worn dj. **$15.00\£8.00** *(CB 1/90)*

Brookes, John. Room Outside. NY, 1970. 192pp. VG in dj. **$15.00\£8.00** *(RW 11/86)*
Another copy. NY, 1970. 2nd prntng. 192pp. VG in dj. **$15.00\£8.00** *(CB 12/89)*

Brooklyn Botanic Garden Record. The Rose Garden of the Brooklyn Botanic Garden. Guide #9. Vol XXVIII, no. 3, July 1939. 159/215pp. VG in aging wrps. **$24.00\£13.00** *(Hrt 12/89)*

Brooks, E. St. John. Sir Hans Sloane. London, 1954. 234pp. VG in dj. **$35.00\£19.00** *(Hrt 12/89)*

Brooks, Reid M. & H.P. Olmo. Register of New Fruit and Nut Varieties. Berkeley, 1972. 2nd edition. 708pp. Extremities soiled, light shelf wear. **$22.00\£12.00** *(BB 12/88)*

Brooks, Sarah Warner. A Garden with House Attached. Boston, 1910. 118pp. VG. **$15.00\£8.00** *(AB 2/89)*

Brotherston, R.P. The Book of the Carnation. London, 1914. 95pp. G overall. **$10.00\£6.00** *(CB 1/90)*

Broun, Maurice. Index to North American Ferns. Publ. by compiler. Orleans, 1938. 217pp. F. **$17.50\£10.00** *(WH 1/90)*

Brown, Alice Cooke. Early American Herb Recipes. NY, 1966. 152pp. VG in dj. **$22.00\£12.00** *(Hrt 12/89)*

Brown, B. Frank. Florida's Beautiful Crotons. Indiatlantic, 1960. 143pp. F in dj. **$22.00\£12.00** *(BB 12/88)*

Brown, B.S. Modern Propagation of Tree Fruit. NY, 1916. 1st ed. 1st 1000, [xi], 174pp. VG. **$22.00\£12.00** *(Hrt 12/89)*

Brown, C. Barrington & William Lidstone. Fifteen Thousand Miles on the Amazon and Its Tributaries. London, 1878. xiii, 520pp. Glt lettered spine. Dec brds. VG. **$160.00\£89.00** *(Hrt 12/89)*

Brown, C.A. Louisiana Trees and Shrubs. 1972. Reprnt. VG. **$23.00\£12.50 DL 89/90)**

Brown, Glen. European and Japanese Gardens. Phil, 1902. 162pp. Tender spine. **$35.00\£19.00** *(Hrt 12/89)*

Brown, Glenn. European and Japanese Gardens. Phil., 1902. 162pp. 4to, brown cloth, moderate wear. **$40.00\£22.00** *(MD 7/89)*

Brown, H.P. Trees of New York State, Native and Naturalized. Syracuse, 1922. 401pp. Gilt embossed green cloth. VG. **$27.50\£15.00** *(BB 12/88)*

Brown, J. et al. Succulents for the Amateur. 167pp. Abbey, 1939. Spine rubbed else VG. **$10.00\£6.00** *(GW 12/88)*

Bruce, "Harold". Gardens of Winterthur in All Seasons. NY, 1968. 196pp. Dj torn and cloth shelf worn. Contents VG. **$20.00\£11.00** *(AB 4/89)*
Another copy. Viking, 1968. 196pp. VG in dj. **$16.00\£9.00** *(GW 8/89)*

Bruce, Harold. The Gardens of Winterthur in All Seasons. NY, 1968. 196pp. Folio. F in slightly trn dj. **$20.00\£11.00** *(BF 2/90)*

Buel, J. Farmer's Instructor. NY, 1854. 2 vols. Front covers loose, contents a little stained. **$40.00\£22.00** *(AB 2/89)*

Buist, Robert. American Flower-Garden Directory. Phila., 1851. 4th ed. 339pp. Orig cloth with new spine. Inner hinges strengthened else G. **$65.00\£36.00** *(ABx 3/89)*
Another copy. Phila., 1852. 5th ed. [xvi], 13-339pp. Light foxing else near F. **$60.00\£33.00** *(Hrt 12/89)*

Buist, Robert. Family Kitchen Gardener. NY, 1856. 216pp. Tan cloth, spine worn. Contents VG. **$40.00\£22.00** *(AB 2/89)*

Bunbury, Charles J.F. Journal of a Residence at the Cape of Good Hope. NY, 1969. [xii], 300pp. As New. **$40.00\£22.00** *(JI 12/88)*

Bunyard, Edward A. Old Garden Roses. London, 1936. [xii], 163pp. Glt stmpd clth, very nice. **$65.00\£36.00** *(Hrt 12/89)*

Bunyard, Edward. Old Garden Roses. London, 1936. 1st ed. 163pp. G. **$81.00\£45.00** *(ABx 3/89)*

Bunyard, G. & O. Thomas. The Fruit Garden. 1904. 1st ed. Orig clth. Well preservd cpy. Scarce. **$63.00\£35.00** *(DL 89/90)*

Bunyard, G. Apples & Pears. 1911. 1st ed. Becoming scarce. VG. **$18.00\£10.00** *(DL 89/90)*

Bunyard, George and Owen Thomas. The Fruit Garden. London, 1904. [xviii], 507pp. 4to. Moderate foxing else VG. **$69.00\£38.00** *(BB 12/88)*

(Burbank). Luther Burbank: His Methods and Discoveries and Their Practical Application. NY, 1914-15. 12 vols. 1 plt mssng else complete. Some pencil notes. T.e.g. **$150.00\£83.00** *(RW 6/89)*

Burbank, Luther. How Plants Are Trained to Work for Man. NY, 1921. 8 vols. G. **$81.00\£45.00** *(ABx 3/89)*

Burbidge, F.W. Cultivated Plants, Their Propagation and Improvement. Edinburgh, 1877. 618pp. Ownr's sig else VG. **$60.00\£33.00** *(BF 2/90)*

Burbidge, F.W. Domestic Floriculture – Window Gardening and Floral Decorations. . . Culture and Arrangement of Plants and Flowers as Domestic Ornaments. 2nd and last ed. 1875. Scarce. Well preserved cpy. **$63.00\£35.00** *(DL 89/90)*

Burbridge, F.W. Cool Orchids and How to Grow Them. London, 1874. 1st ed. [viii], 160pp. 12mo. Cover sl rubbed with worn corners. **$155.00\£86.00** *(JI 12/88)*

Burbridge, F.W. The Garden of the Sun. London, 1880. [xviii], 364pp. 24pp advts. Mssng f.f.e.p. & sl tendr else VG in dec brds. **$90.00\£50.00** *(Hrt 12/89)*

Buret, F. Le Champignon. Paris, 1925. [x], 228pp. Wrpprs. G. **$27.00\£15.00** *(WW 1/90)*

Burgess, E. Species and Behavior of Biotian Asters; Pre-Clusian Botany in Its Relation to Aster. Torrey Botanical Society, 1906. 866pp. A few library stamps else VG. **$25.00\£14.00** *(GW 8/89)*

Burgess, Edward. History of Pre-Clusian Botany in Its Relation to Aster, NY. Torrey Botanical Club, 1902. 447pp. Spine peeling, wraps off. **$45.00\£25.00** *(AB 4/89)*

Burgess, Thornton W. The Burgess Flower Book for Children. NY, 1929. [xviii], 350pp. VG. **$25.00\£14.00** *(ABx 3/89)*

Bush-Brown, Louise & James. Portraits of Philadelphia Gardens. Phil., 1929. 179pp. Clth bcked brds, signed on verso by Louise Bush-Brown. VG. **$65.00\£36.00** *(MD 2/90)*

Bush-Brown, Louise. Flowers for Every Garden. Boston, 1927. 194pp. Ownr's b.p. else VG. **$12.00\£7.00** *(BF 2/90)*

Bush-Brown, Louise. Flowers for Every Garden. Boston, 1927. 1st ed. 194pp. Decorative cover. **$12.50\£7.00** *(RW 11/86)*

Bush-Brown, Louise. Garden Blocks for Urban America. NY, 1969. 126pp. VG in worn dj. **$10.00\£6.00** *(RW 6/89)*

Butler, E.J. & S.G. Jones. Plant Pathology. London, 1949. [xii], 979pp. Ry 8vo. Lettering on spine faded. **$36.00\£20.00** *(WW 1/90)*

Butler, June Rainsford. Floralia-Garden Paths and By-Paths of the 18th Century. Chapel Hill. U of N.C. Press, 1938. Ed of 500. 187pp. Spine lghtly sunned else F. **$55.00\£31.00** *(BF 2/90)*

Bylov, V.N. et al. Rozy. Moscow, 1962. Lge 8vo, 224pp. In Russian. VG. **$55.00\£31.00** *(PBx 11/89)*

Byne, Arthur & Mildred Stapley. Rejeria of the Spanish Renaissance. NY. Hispanic Soc. Publ. #87. 1914. [vii], 103pp. Folio. T.e.g. orig clth, rbbd & sl tender. **$60.00\£33.00** *(Hrt 12/89)*

Byne, Mildred Stapley & Arthur Byne. Spanish Gardens and Patios. NY, 1924. 305pp. (3) advts. Orig gilt pict clth. VG. **$135.00\£75.00** *(Hrt 12/89)*
Another copy. NY, 1928. 305pp. Ownr's inscrp, faint wtrstn to bottom edges causing rippling to pp. Else near VG in rbbd and slightly wtr stned pict clth. **$100.00\£56.00** *(BF 2/90)*

C

Cabot, Joseph. An Innocent in the Garden. London, 1954. 1st ed. 205pp. VG in soiled, trn dj. **$15.00\£8.00** *(CB 1/90)*

Cadbury, D.A., J.G. Hawkes and R.C. Readett. A Computer Mapped Flora of Warwickshire. London, 1971. [ix], 768pp. 4to. Clth. G. **$54.00\£30.00** *(WW 1/90)*

Caledonian Horticultural Society; Memoirs. Edinburgh, 1819. 2nd ed. Vol. I. 475pp. Mdrn hlf clf w/marbld brds. F. **$50.00\£28.00** *(PBx 11/89)*

Calthrop, Dion Clayton. A Diary of an Eighteenth Century Garden. NY, nd. (@1927). 128pp. VG in tattered dj. **$16.00\£9.00** *(CB 8/89)*

Calthrop, Dion. The Charm of Gardens. Black. London, 1910. [xii], 240pp + 4pp of ads. 32 full-page illus. in color. T.e.g., dec cvr. VG. **$30.00\£17.00** *(RW 11/86)*

Cameron, Katharine. The Flowers I Love. London & Edinburgh, nd. (@1916). [xi], 87pp. G. **$72.00\£40.00** *(ABx 3/89)*

Camp, W.H., V.R. Boswell, J.R. Magness. The World in Your Garden. Wash., 1957. 231pp. Ink insc. on e.p., dj chppd. **$15.00\£9.00** *(RS 2/89)*

Camp, Wendell H. et al. The World in Your Garden. Wash., 1957. 231pp. VG. **$5.00\£3.00** *(WH 1/90)*

Campbell, A.J. Golden Wattle, Our National Floral Emblem. Melbourne, 1921. 63pp. Clth-bckd brds. G. **$36.00\£20.00** *(WW 1/90)*

(Canada Forestry Branch) Canadian Woods, Their Properties & Uses. 1951. 2nd ed. 368pp. G. **$20.00\£11.00** *(PBx 11/89)*

Campbell, D. An Outline of Plant Geography. McMillan, 1926. 392pp. Front hinge cracking else VG. **$12.00\£7.00** *(GW 12/88)*

Candolle, A.P. de and K. Sprengel. Elements of the Philosophy of Plants. Edinburgh, 1821. [xxxii], 486pp. Hlf calf, slightly rbbd. **$108.00\£60.00** *(WW 1/90)*

Cane, P.S. Garden Design Quarterly Journal, nos. 22-29, 1935-37. 8 prts as issued. VG. **$57.00\£32.00** *(DL 89/90)*

Cane, Percy. Modern Gardens, British and Foreign. London, 1927. Folio. 166pp. VG. **$55.00\£31.00** *(BF 2/90)*
Another copy. 1926-27. Becoming scarce. VG. **$31.00\£17.50 DL 89/90)**

Cane, Percy. The Earth Is My Canvas. London, 1956. [ix], 161pp. VG. **$35.00\£19.00** *(Hrt 12/89)*
Another copy. 1956. VG in dj. **$45.00\£25.00** *(DL 89/90)*

Cannon, W. General & Physiological Features of the Vegetation of S. Africa. Carnegie, 1924. 159pp. Clean x-lib. **$25.00\£14.00** *(GW 8/89)*

Carleton, Mark. Small Grains. NY, 1920. 699pp. Light wear, contents VG. **$15.00\£8.00** *(AB 2/89)*

Carr, Rachel E. Stepping Stones to Japanese Floral Art. Tokyo, 1957. 3rd prntng. 109pp. Bnd accordian-style in green silk, slightly rbbd, with silk ties. **$20.00\£11.00** *(BF 2/90)*

Carter, Annie Burnham. In an Herb Garden. NJ, 1947. 209pp. VG in dj. **$7.50\£4.00** *(RW 11/86)*

Cartwright, Julia. Italian Gardens of the Renaissance and Other Studies. London, 1914. [xii], 298pp. Ink sig & sl tender. **$45.00\£25.00** *(Hrt 12/89)*

Carver, Humphrey. Compassionate Landscape. Toronto, 1975. [vii, ii], 251pp. As New in dj. **$20.00\£11.00** *(Hrt 12/89)*

Cary, Merritt. Life Zone Investigations in Wyoming. Wash., GPO, 1917. 94pp. Wraps. VG. **$15.00\£8.00** *(AB 4/89)*

Castle, Agens & Egerton. Our Sentimental Garden. Phil., 1914. 305pp. VG. **$25.00\£14.00** *(RW 11/86)*

Castle, Lewis. Orchids. London, 1887. 3rd ed. 129pp + ads. G. **$34.00\£19.00** *(ABx 3/89)*

Cautley, Marjorie. Garden Design, Principles of Abstract Design. NY, 1935. 312pp. Cloth lightly worn, contents VG. **$35.00\£19.00** *(AB 4/89)*

Cautley, Marjorie. Garden Design. NY, 1935. 312pp. Ex-priv libry-env & sig, in orig gilt brds. Sl tender else VG. **$35.00\£19.00** *(Hrt 12/90)*

Cavendish, Marshall. Illustrated Encyclopedia of Gardening, 1966-70. 8 vols, plastic folders. 96 prts as issued. Well preservd. **$144.00\£80.00** *(DL 89/90)*

Cecil, Mrs. Evelyn. London Parks and Gardens. NY, 1907. 384pp. Decorated cloth, spine faded else G. **$108.00\£60.00** *(ABx 3/89)*

Cecil, The Hon. Mrs. Evelyn (Alicia Amherst). A History of Gardening in England. London, 1910. [xviii], 393pp. 3rd enlrgd ed. Green cloth, neat x-lib. Tight. **$68.00\£38.00** *(BB 12/88)*

Cecil, The Hon. Mrs. Evelyn (Alicia Amherst). Children's Gardens. London, 1902. 1st ed. 212pp. Dec cvr, t.e.g. **$25.00\£14.00** *(RW 11/86)*

Cecil, the Hon. Mrs. Evelyn. A History of Gardening in England. NY, 1910. 3rd ed. Enlrgd. 393pp. Scttrd fxng, hinges crckd, bndng rbbd with tears at spine edges, but solid cpy. **$95.00\£53.00** *(BF 2/90)*

Cesi, F. Phytosophicarum Tabularum. Rome, 1904. [xiv, viii], 85pp. New brds, preserving part of orig sides. **$45.00\£25.00** *(WW 1/90)*

Chadbund, Geoffrey. Flowering Cherries. London, 1972. 160pp. G in dj. **$29.00\£16.00** *(ABx 3/89)*

Chamberlain, C.J. Methods in Plant Histology. Chicago, 1932. 5th rev. ed. G. **$27.00\£15.00** *(WW 1/90)*

Chandler, Katherine. Habits of California Plants. SF, 1903. 204pp. Near F. **$36.00\£20.00** *(Hrt 12/89)*

Chapple, F.J. The Heather Garden. London, 1952. 1st ed. 180pp. In chppd dj. **$25.00\£14.00** *(RS 2/89)*

Chapple, Fred. The Heather Garden. London, 1952. 180pp. G in dj. **$27.00\£15.00** *(ABx 3/89)*

Chase, Agnes. First Book of Grasses. San Antonio, 1937. [xiii], 125pp. Rev. ed. VG. **$17.50\£10.00** *(WH 1/90)*

Chase, J. Smeaton. Cone-Bearing Trees of the California Mountains. Chicago, 1911. 99pp. G. **$23.50\£13.00** *(ABx 3/89)*

Chaumeton, F.P., J.B. Chamberet & J.L.M. Poiret. Flore Medicale. Paris, 1814-20. 7 vols in 8. Contemp hlf clf, gilt (trifle wrn). G cpy; some usual fxng and brwning of paper. **$5940.00\£3300.00** *(WW 1/90)*

Chevalier, A. Les Iles du Cap Vert: Flore de l'Archipel. Paris, 1935. Tome XV. 733-1090pp. Wrpprs, neatly reprd. **$63.00\£35.00** *(WW 1/90)*

Cheyney, E.G. What Tree Is That? NY, 1927. [xvi], 185pp. Small nick to fore-edge else bright. **$11.50\£6.00** *(BB 12/88)*

Chickering, Carol. Flowers of Guatemala. Norman, 1973. 132pp. 4to. VG in dj. **$15.00\£8.00** *(BB 12/88)*

Chittenden, F.J. (ed). R.H.S. Dictionary of Gardening & Supplement. London, 1951, 1956. 5 vols together. 2332pp, [vii], 334pp. Slightly used. **$234.00\£130.00** *(WW 1/90)*

Cholnoky, B.J. Die Okologie der Diatomeen in Binnengewaessern. Lehre, 1968. 699pp. Ry 8vo. Clth. G. **$90.00\£50.00** *(WW 1/90)*

Christchurch Botanic Gardens 1863-1963. N.Z. Christchurch, 1963. 182pp. G. **$27.00\£15.00** *(WW 1/90)*

Christensen, Clyde. Common Edible Mushrooms. Minneapolis, 1969. 3rd prtng. [x], 114pp. VG. **$11.00\£6.00** *(CB 12/89)*

Christensen, Clyde. The Molds & Man. Mnpls, 1951. 244pp. VG in fair dj. **$5.00\£3.00** *(WH 1/90)*

Christian, Frances A. & Susanne W. Massie (eds). Homes and Gardens in Old Virginia. Richmond, 1931. 368pp. About VG. **$30.00\£17.00** *(AT 3/89)*

Church, A.H. Introduction to the Plant Life of the Oxford District. London, 1922-25. 3 prts, wrappers. Oxford Botanical Memoirs 13, 14, 15. G. **$27.00\£15.00** *(WW 1/90)*

Church, E.R. The Home Garden. NY, 1881. 121pp. Orig clth, wrn at crnrs; soiled; spine drknd, rear e.p. mssng. **$20.00\£11.00** *(RS 2/89)*

Church, Thomas. Gardens Are for People. NY, 1955. 248pp. Sm folio. VG in wrn dj. **$65.00\£36.00** *(Hrt 12/89)*

Clair, Colin. Of Herbs and Spices. London, 1961. 276pp. VG in dj. **$7.50\£4.00** *(RW 11/86)*

Clapham, Sidney. Primulas. S. Brunswick, 1972. 143pp. Fine in dj. **$17.50\£10.00** *(BB 12/88)*

Clark, H.F. The English Landscape Garden. 1948. 1st and only ed. Scarce. VG in dj. **$23.00\£12.50** *(DL 89/90)*

Another copy. London, 1948. 64pp. VG. $30.00\£17.00 *(BF 2/90)*

Clark, H.F. The English Landscape Garden. London, 1948. [xi], 64pp + 56 photo plates. VG. **$15.00\£8.00** *(RW 11/86)*

Clark, William. Gardening for Color. Boston, 1954. [viii], 212pp. VG in older dj. **$15.00\£8.00** *(Hrt 12/89)*

Clarke, C.B. Commelynaceae et Cyrtandraceae Bengalenses. Calcutta, 1874. [ii], 133pp. Folio. Clth-bcked brds. Some heavy damp stning in lower mrgns. **$180.00\£100.00** *(WW 1/90)*

Clarke, F. The Genera of Fungi. Wilson, 1931. 227pp. VG. **$35.00\£19.00** *(GW 8/89)*

Clarke, H. Harold. Small Fruits for Your Home Garden. NY, 1958. [xiii], 372pp. VG in dj. **$7.50\£4.00** *(RW 11/86)*

Clarke, J. Harold. Small Fruits for Your Home Gardens. NY, 1958. [xiii], 372pp. VG in dj. **$7.50\£5.00** *(RW 6/89)*

Clarke, J.H. (ed). Rhododendrons for Your Garden. Amer. Rhod. Soc., 1961. G. **$16.00\£9.00** *(PBx 11/89)*

Clarke, M.U. Nature's Own Gardens. London, 1907. [xii], (1), 280pp. Orig decor clth, VG. **$50.00\£28.00** *(RS 2/89)*

Clarkson, L. Violet among the Lilies. NY, 1885. A.e.g., pict. gold & silver stmped cloth shows some wear else quite G. **$80.00\£44.00** *(Hrt 12/89)*

Clarkson, R.E. Green Enchantment. NY, 1940. 1st ed. [xix], 382pp. In chppd dj. **$27.00\£15.00** *(RS 2/89)*

Clarkson, Rosetta E. Herbs, Their Culture and Uses. NY, 1966. [vi], 226pp. VG in dj. **$20.00\£11.00** *(Hrt 12/89)*

Clarkson, Rosetta E. Magic Fragrance. Ltd. ed. New Rochelle, 1937. Spine fade else VG. **$75.00\£42.00** *(Hrt 12/89)*

Clarkson, Rosetta. Magic Gardens. NY, 1939. [xviii], 369pp. Near F. **$45.00\£25.00** *(Hrt 12/89)*

Clausen, Robert T. Sedum of North America North of the Mexican Plateau. Ithaca, 1975. 742pp. F in dj. **$47.50\£26.00** *(WH 1/90)*
Another copy. Ithaca, 1975. 742pp. As New in dj. **$50.00\£28.00** *(AB 4/89)*

Clay, S. The Present Day Rock Garden. London, 1937. [xxvi], 681pp. VG in chppd dj. **$85.00\£47.00** *(RS 2/89)*

Cleaveland, N. The Rural Cemeteries of America, Green-wood. NY, 1847. 1st ed. A.e.g. [viii], 94pp. View of entrance listed for pg 8 is on title pg. Some shlfwr to spine, slght tender & some lght fxng to a few plts else about F. Gilt stmpd dec brds. **$145.00\£81.00** *(Hrt 12/89)*

Clements, Edith. Flowers of Coast and Sierra. Wilson, 1928. 226pp. VG. **$22.50\£13.00** *(GW 12/88)*

Clements, F. The Genera of Fungi. Wilson, 1909. 227pp. VG. **$35.00\£19.00** *(GW 12/88)*

Clements, Frederic E. & Edith S. Rocky Mountain Flowers. Wilson, NY, 1914. 392pp. Scarce. VG. **$30.00\£17.00** *(WH 1/90)*

Clements, Frederic E. Plant Succession and Indicators. NY, 1928. 453pp. F. **$42.50\£24.00** *(WH 1/90)*

Clements, Julia. Flower Arrangements in Stately Homes. London, 1966. 124pp. Inscribed. VG in worn dj. **$30.00\£17.00** *(RW 6/89)*

Clute, Willard N. Our Ferns in Their Haunts. NY, 1901. [xii], 332pp. VG. **$25.00\£14.00** *(RW 11/86)*

Clute, Willard N. Swamp and Dune. Indianapolis, 1931. 92pp. Glt stmpd clth. F. **$21.00\£12.00** *(Hrt 12/89)*

Clute, Willard N. The Fern Allies of North America North of Mexico. NY, 1905. [xiv], 278pp. VG. **$25.00\£14.00** *(RW 11/86)*

Clutton-Brock, A. Studies in Gardening. NY, 1916. [xxxviii], 337pp. Spine faded; lght wr. **$20.00\£11.00** *(RS 2/89)*

Coats, Alice M. Flowers and Their Histories. NY, 1971. 1st Amer. ed. 346pp. F in dj. **$30.00\£17.00** *(BF 2/90)*

Coats, Alice M. The Book of Flowers—Four Centuries of Flower Illustration. 2nd ed. 1974. As new in dj. **$72.00\£40.00** *(DL 89/90)*

Coats, Alice M. The Treasury of Flowers. NY, 1975. 121pp. VG in dj. **$15.00\£8.00** *(RW 11/86)*
Another copy. McGraw-Hill, nd. 30pp. VG in dj. **$25.00\£14.00** *(JI 7/88)*

Coats, Alice. The Plant Hunters. NY, 1969. 400pp. F in dj. **$25.00\£14.00** *(WH 1/90)*

Coats, Peter. Flowers in History. NY, 1970. 1st Amer. ed. 264pp. F in dj. **$50.00\£28.00** *(BF 2/90)*

Coats, Peter. Great Gardens of Britain. London, 1967. 1st ed. 288pp. VG in torn dj. **$43.00\£24.00** *(ABx 3/89)*

Coats, Peter. Great Gardens of Britain. London, 1967. 288pp. VG in reprd dj. **$45.00\£25.00** *(CB 1/90)*
Another copy. London, 1967. Folio. VG in wrn, chppd dj. **$50.00\£28.00** *(BF 2/90)*

Coats, Peter. Great Gardens of the Western World. NY, 1963. 288pp. 4to. 40 color plates, 350 monochrome. VG in dj. **$45.00\£25.00** *(RW 11/86)*
Another copy. NY, 1963. 1st ed. 288pp. VG in bright dj. Spine rubbed. **$45.00\£25.00** *(CB 12/89)*

Coats, Peter. Roses. NY, 1962. 128pp. Boxed. VG. **$15.00\£8.00** *(RW 6/89)*
Another copy. 1962. G. **$20.00\£11.00** *(PBx 11/89)*

Cobbett, William. A Year's Residence in the United States. NY, 1819. 1st US in 3 vols. 432pp. Uniform dmp stn vol. I & II not materially affecting text. Slightly vol. III, aged 1/2 leather on brds. Despite discoloration, quite a nice set. **$180.00\£100.00** *(Hrt 12/89)*

Cobbett, William. The English Gardener. London, 1838. First published 1829. 338pp. Well bound in recent brown quarter calf, marbled boards. **$117.00\£65.00** *(ABx 3/89)*

Cobbett, William. The Woodlands. London, 1825. Prnted & publ by author. 1825. Unpaginated, 342pp. 2pp advs. 1/2 calf, marbled brds. Rubbing & some scattered fxing. **$125.00\£69.00** *(Hrt 12/89)*

Cobbett, Wm. The American Gardener. Man. Co. Claremont, NH, nd. (circa 1830s). 230pp. Blindstamp cloth, very light wear. **$45.00\£25.00** *(AB 2/89)*
Another copy. Concord, NH, 1842. 1st Amer. ed. 271pp. Leather spine well worn. Contents VG. **$60.00\£33.00** *(AB 2/89)*

Cochet-Cochet & Mottet, S. Les Rosiers. 5th ed. 1925. Sft cvr. 379pp. Copy of J.H. Nicolas. G. **$25.00\£14.00** *(PBx 11/89)*

Cockayne, L. & E. Phillips Turner. The Trees of New Zealand. Wellington, 1958. 4th ed rev., 1928. 182pp. VG in chipped dj. **$22.00\£12.00** *(BB 12/88)*

Cockayne, L. New Zealand Plants and Their Story. Wellington, 1967. 4th edition. 269pp. Illus. cloth, corners bumped. Light page wrinkling. **$35.00\£19.00** *(BB 12/88)*

Coffin, David R. (ed). Italian Gardens. 1st Dumbarton Oaks Colloquium on the History of Landscape Architecture. WA, 1972. [viii], (1) contents, (1) blank, 114pp. G. **$24.00\£13.00** *(Hrt 12/89)*

Coffman, F.A. (ed). Oats & Oat Improvement. 1961. 650pp. Unused in shlfwrn dj. **$25.00\£14.00** *(PBx 11/89)*

Coker, William Chambers & John Nathaniel Couch. The Gasteromycetes of the Eastern United States and Canada. NY, 1974. [vi], 284pp. Wraps. VG. **$20.00\£11.00** *(JI 12/88)*

Coley, Hilda. Our Heritage of Garden Flowers. London, 1936. 86pp. G. **$11.00\£6.00** *(ABx 3/89)*

Colin, Jane. Herbs & Spices for Health and Beauty. London, 1962. 96pp. VG in dj. **$15.00\£8.00** *(AB 4/89)*

Colman, E.A. Vegetation & Watershed Management. NY, 1953. 412pp. F in dj. **$15.00\£8.00** *(WH 1/90)*

Colman, Pamela. The Ladies' Vase of Wild Flowers: A Collection of Gems from the Best Authors. New Orleans, 1854. 126pp. of flower poetry, 1 color plt. Variant imprint. Brwn clth, glt & blind stmpd, a.e.g. VG. **$35.00\£19.00** *(BF 2/90)*

Colonna, Rabio. Lyncei Oytobaenoe Cvi Accessit Vita Fabi et Lynceorum Notitia. Milan, 1744. 2nd ed. [lii], 134pp. Recent full leather, near F. **$950.00\£528.00** *(Hrt 12/89)*

Comstock, F.G. A Practical Treatise on the Culture of Silk. Hartford, U.S., 1839. 2nd ed. Revised & improved. 96pp. Original printed card covers, (silk?) spine, this frayed. Else in excellent condition. Rare. **$77.50\£43.00** *(ABx 3/89)*

Conard, Henry S. & Henri Hus. Water-Lilies. Garden City & NY, 1914. Small 8vo. [xiii], 228pp. G. **$20.00\£11.00** *(CB 8/89)*

Conard, Henry. The Water Lilies. WA. Carnegie Institution, 1905. 279pp. Half brown morocco. Rare. G. **$216.00\£120.00** *(ABx 3/89)*

Conrad, Henry S. & Henry Hus. Water Lilies and How to Grow Them. NY, 1907. [xiii], 228pp. Glt cvrd brds. VG. **$35.00\£19.00** *(Hrt 12/89)*
Another copy. Garden City & NY, 1914. [xiii], 228pp. Lght staining upper crnr. G. **$20.00\£11.00** *(CB 1/90)*

Cook, E.T. (ed). The Century Book of Gardening. London. Country Life, nd. [ix], 610, [xiii]pp. 4to. Illus. a.e.g. dec cvr. Some wear to spine else VG. **$35.00\£19.00** *(RW 11/86)*
Another copy. Nd (1900s). 610, [xiii]pp. Decorated cloth cover. G. **$54.00\£30.00** *(ABx 3/89)*

Cook, E.T. Gardening for Beginners. London, 1901. [vii], 495pp. Decorated cloth. VG. **$35.00\£19.00** *(RW 11/86)*

Cook, E.T. Gardens of England. London, 1911. [ix], 199pp. Dec cvr. Sl chip on top of spine. VG. **$20.00\£11.00** *(RW 11/86)*

Cook, E.T. The Century Book of Gardening. London, nd, c. 1900. [xiii], 610pp. Last few pp wtrstnd at top, hinges crckd, large tear across spine. G only. **$30.00\£17.00** *(BF 2/90)*

Cook, Mrs. Harriet N. The Trees, Fruits & Flowers of the Bible. NY, 1846. 12mo. Gilt blind stamping. Not perfect but attractive. **$26.00\£14.00** *(PBx 11/89)*

Cook, O.F. & C. Doyle. Three New Genera of Stilt Palms from Colombia. Wash., GPO, 1913. USNH. 20pp. Wraps light wear. **$15.00\£8.00** *(AB 4/89)*

Cook, O.F. History of the Coconut Palm in America. Washington, 1910. [vii], 271-342pp, [ix-xiii]. USNH, vol. 14, part 2. Wrappers. **$20.00\£11.00** *(BB 12/88)*

Cook, O.F. Origin and Distribution of Cocoa Palm. Wash., GPO, 1901. USNH. 37pp. VG. **$15.00\£8.00** *(AB 4/89)*

Cook, O.F. Relationships of the False Date Palm of Florida Keys. Wash., GPO, 1913. USNH. 18pp. VG. **$10.00\£6.00** *(AB 4/89)*

Cook, T.H., James Douglas & J.F. McLeod. Carnations and Pinks. NY, nd. 116pp. VG. **$10.00\£6.00** *(RW 11/86)*

Cooke, A.O. A Book of Dovecotes. London, 1920. 286pp + index. Grey cloth spine & paper covered boards. Corners slightly worn. G. **$99.00\£55.00** *(ABx 3/89)*

Cooke, M.C. Fungi. NY, 1875. 1st US ed. [xxii], 312pp. VG. **$45.00\£25.00** *(JI 12/88)*

Coombs, Sarah V. South African Plants. NY, 1936. [xvi], 364pp. G. **$38.00\£21.00** *(Hrt 12/89)*

Coon, N. Practical Violet Culture. NY, 1925. 127pp. Lght wr, soiling; hinge splttng. **$12.50\£7.00** *(RS 2/89)*

Cooney, Loraine M. Garden History of Georgia 1733-1933. Atlanta, 1933. 4to, red cloth with paper label. 458pp.

#115 from an edition of 1500. Lacks front end paper, about 4" of inner margin of title page is split, minor foxing to title page. **$125.00\£69.00** *(MD 7/89)*

Cooper, Elizabeth. My Lady of the Chinese Courtyard. NY, 1914. 9th prtng. 262pp. VG in gilt/photo cvr. **$30.00\£17.00** *(Hrt 12/89)*

Copeland, Morris. Country Life: A Handbook of Agriculture, Horticulture, and Landscape Gardening. Boston, 1860. 2nd ed. [x], 814pp. Spine very tender and faded, wrn brds else VG internally. **$85.00\£47.00** *(Hrt 12/89)*

Copeland, R. Morris. Country Life. Boston, 1859. Brown cloth, somewhat spotted. 813pp. Shelf-worn, extremities worn. **$100.00\£56.00** *(MD 11/88)*

Cornell, Ralph. Conspicuous California Plants. Pasadena, 1938. 1st edition. 192pp. Ltd. ed, 1500 copies. Decorated brick colored cloth. Bookplate of Norman Field Sisson. Edges browning, else VG. **$55.00\£31.00** *(BB 12/88)*

Corner, E.J.H. Wayside Trees of Malaya. 1952. 2nd ed. 2 vols. Nice, clean set. **$135.00\£75.00** *(DL 89/90)*

Correvon, H. & P. Robert. The Alpine Flora. Geneva, 1912. 436pp. G. **$45.00\£25.00** *(WW 1/90)*

Correvon, H. Album des Orchidées. Geneva and Paris, 1899. [xv], 92pp. Clth. bookplate of/ present. inscrip signed by Ellen Willmott. Faint waterstn extreme crner of some plts, a little fxng. G. **$216.00\£120.00** *(WW 1/90)*

Correvon, H. Rock Gardening and Alpine Plants. NY, 1930. [xiv], (2), 544pp. Lght wr. **$27.00\£15.00** *(RS 2/89)*

Correvon, H. Rock Gardens and Alpine Plants. NY, 1930. [xv], 544pp. Orig clth, trifle used. **$45.00\£25.00** *(WW 1/90)*

Correvon, Henry. Rock Garden and Alpine Plants. NY, 1930. [xiv], 544pp. VG. **$20.00\£11.00** *(RW 11/86)*
Another copy. NY, 1930. 544pp. Small repair to top of front hinge. G. **$54.00\£30.00** *(ABx 3/89)*

Cory, V. Catalogue of the Flora of Texas. TX, 1937. 130pp. Worn, soiled wraps. **$11.00\£6.00** *(GW 12/88)*

Costa, B.C. Cincinnato da and Luiz de Castro. Le Portugal au Point de Vue Agricole. Lisbonne, 1900. [xxxvii] + 965pp. Wrps bound in. G. **$45.00\£25.00** *(WH 1/90)*

Coulter, John & J. Rose. Monograph of North American Umbelliferae. Wash., GPO, 1900. USNH. Vol. VII, #1. 256pp. Wraps. VG. **$22.00\£12.00** *(AB 4/89)*

Country Life Library. Gardens Old and New. London, nd (1900). 1st ed. [xxiv], 295pp. Decorated cloth covers. G. **$126.00\£70.00** *(ABx 3/89)*

Cowan, May. Inverewe. London, 1964. 152pp. G in dj. **$21.50\£12.00** *(ABx 3/89)*

Cowan, R. Flora Newotropica: Swartzia. Hafner, 1967. 227pp. Wrs. **$16.00\£9.00** *(GW 8/89)*

Cowen, D. Flowering Trees and Shrubs in India. Thacker, 1957. 142pp. 3rd ed. VG. **$40.00\£22.00** *(GW 8/89)*

Cox, E.H.M. & G.C. Taylor. Primulas for Garden & Greenhouse. Oxford, 1947 (1928). [vii], 86pp. G in chipped dj. **$12.00\£7.00** *(BB 12/88)*

Cox, E.H.M. A History of Gardening in Scotland. London, 1935. 1st (& only) ed. 229pp. G. **$45.00\£25.00** *(ABx 3/89)*

Cox, E.H.M. Farrer's Last Journey. London, 1926. 244pp. Lght fxng to edges else VG in faded, sl trn dj. Scarce. **$145.00\£81.00** *(Hrt 12/89)*

Cox, E.H.M. Plant Hunting in China. London, 1945. 1st ed. 230pp. G in slightly torn dj. **$27.00\£15.00** *(ABx 3/89)*
Another copy. Collins, Scientific Book Guild. 1945. 230pp. **$35.00\£19.00** *(GW 12/88)*

Cox, E.H.M. Plant Hunting in China. London, 1945. 230pp. Spine faded, frnt jnt slightly loose. **$45.00\£25.00** *(WW 1/90)*
Another copy. London, 1945. 230pp. Quite G in dj. **$50.00\£28.00** *(Hrt 12/89)*

Cox, E.H.M. Rhododendrons for Amateurs. London, 1924. [xvi], 112pp. Some sl brwning & fading to brds else VG. **$30.00\£17.00** *(Hrt 12/89)*
Another copy. London, 1924. [xv], 112pp. Name f.f.e.p. else VG. **$22.00\£12.00** *(CB 1/90)*

Coxe, William. A View of the Cultivation of Fruit Trees and the Management of Orchards and Cider. . . Accurate Descriptions of Apples, Pears, Peaches, Plums, Cherries, Apricots, Quince. Phila, 1817 (1975). 360pp. Ltd to 500 nmbrd cpies. F in dj. **$32.00\£18.00** *(PBx 11/89)*

Crabb, A.B. Hybrid-Corn Makers. New Bruns., Rutgers, 1948. 331pp. Light wear else VG. Signed by author(?). **$12.00\£7.00** *(AB 4/89)*

Craig, R.T. The Mammillaria Handbook. 1st ed. Pasadena, 1945. [xiv], (2) 390pp. Some wr at extrmities. **$80.00\£44.00** *(RS 2/89)*

Craig, W.N. Lilies & Their Culture in North America. Chicago, 1928. Autographed. Crisp copy. **$20.00\£11.00** *(PBx 11/89)*

Craig, W.N. Lilies and their Culture in North America. Chicago, 1928. 1st ed. 146pp. lght wr. **$20.00\£11.00** *(RS 2/89)*

Cran, M. Hagar's Garden. London, 1941. 285pp. Bmpd crnr; occ pencil mrks. **$20.00\£11.00** *(RS 2/89)*

Cran, Marion. Gardens in America. London, 1931. 1st ed. 320pp. Preliminaries & edges heavily foxed, else VG. **$26.00\£14.00** *(BB 12/88)*

Cran, Marion. Gardens in America. NY, 1932. 320pp. Green clth, paper label. Few water spts to frnt cvr else VG. **$15.00\£8.00** *(CB 1/90)*

Cran, Marion. The Garden of Experience. London, nd. (c1920). 6th printing. 316pp. G. **$11.00\£6.00** *(ABx 3/89)*

Cran, Marion. The Garden of Ignorance. London, 1931. New ed. 268pp. G. **$9.00\£5.00** *(ABx 3/89)*
Another copy. NY, 1913. 273pp, [vii] index. VG. **$25.00\£14.00** *(RW 6/89)*

Cran, Marion. The Story of My Ruin. London, 1924. 320pp. G. **$11.00\£6.00** *(ABx 3/89)*

Crane, Howard H. Gardening on Clay. London, 1963. 120pp. G in dw. **$18.00\£10.00** *(ABx 3/89)*

Crawford, J.H. Wild Flowers of Scotland. London, 1897. 228pp. Decorated cloth. **$27.00\£15.00** *(ABx 3/89)*

Creevey, Caroline. Flowers of Field, Hill & Swamp. NY, 1898. [viii], 568 + 14pp of ads. Decorative cover. **$20.00\£11.00** *(RW 11/86)*

Creevey, Mrs. Caroline. Harper's Guide to Wild Flowers. NY, 1912. 1st ed. [x], 558pp. VG. **$30.00\£17.00** *(JI 4/89)*

Cridland, Robert B. Practical Landscape Gardening. NY, 1929. 3rd ed. 280pp. VG in worn dj. **$15.00\£8.00** *(RW 6/89)*

Cridland, Robert. Practical Landscape Gardening. NY, 1925. 5th prntng. 276pp. Pictorial onlay, gilt stmpng. Extremities wrn, stn on back cvr else VG. **$15.00\£8.00** *(CB 1/90)*

Crisp, Sir Frank. Mediaeval Gardens. London, 1924. Ltd. ed. of 1000 copies, this #368. 2 vols. 140pp + 225 illus.; [xiv] + 314 illus. New end papers. Slight marks on spine. Very handsome decorated cream buckram. VG, bright set. **$432.00\£240.00** *(ABx 3/89)*

Crittenden, Mabel & Dorothy Telfer. Wildflowers of the West. Millbrae, CA, 1975. [viii], 200pp. 8vo. Wraps. As New. **$15.00\£8.00** *(JI 7/88)*

Croasdale, H. Freshwater Algae of Ellesmere Island, NWT. Canada, 1973. 125pp. Wrs. VG. **$12.00\£7.00** *(GW 8/89)*

Crockett, James. Greenhouse Gardening as a Hobby. NY, 1961. 288pp. G in dj. **$12.50\£7.00** *(ABx 3/89)*

Crowe, Sylvia. Garden Design. NY, 1959. 1st Amer ed. 229pp. Near VG in dj. **$20.00\£11.00** *(BF 2/90)*
Another copy. 1958. 1st ed. Difficult to locate. Nice cpy in dj. **$45.00\£25.00** *(DL 89/90)*

Crowe, Sylvia. Garden Design. NY, 1959. 229pp. VG in dj. **$20.00\£11.00** *(RW 6/89)*

Crowell, Joshua Freeman. Garden Wise and Otherwise. Boston, 1934. 153pp. Nr F in older dj. **$15.00\£8.00** *(Hrt 12/89)*

Croy, Mae. Putnam's Garden Handbook. NY, 1917. 359pp. Light wear, internally new. **$10.00\£6.00** *(AB 2/89)*

Cullen, Gordon. The Concise Townscape. London, 1971. 199pp. Wrps VG. **$18.00\£10.00** *(Hrt 12/89)*

Culpeper, N. Culpeper's English Physician and Complete Herbal to Which Are Now First Added... Their Medical & Occult Properties. . . Beautiful & Enriched with Engravings of Upwords of 450 Different Plants... By the late E. Silby, M.D. 14th ed. London, 1810. 2 vols in 1. 656pp. 4to. Fll green glt morocco, dentelles, portrait, 29 col plts (as called for). 13 engrs. in red. Minor imperfections to bndng. Sound tight, clean. **$465.00\£258.00** *(PBx 11/89)*

Culpeper, N. The Complete Herbal; To Which Is Now Added... The English Physician Enlarged. London, 1850. [vi], 398pp. (2), contemp roan. Rbbed, frnt jnt broken. Somewhat fxed and a little soiled. **$135.00\£75.00** *(WW 1/90)*

Culpeper, Nicolas M.D. The Complete Herbal;... with the English Physician Enlarged, and Key to Physic... forming a complete Family Dispensatory and Natural System of Physic. London, 1850. 302pp. Half maroon calf with marbled boards. G. **$153.00\£85.00** *(ABx 3/89)*

Cumming, A. Hardy Chrysanthemums. NY, 1939. [xvi], 168pp. Some brwning of pp; in wrn dj. **$12.00\£7.00** *(RS 2/89)*

Cummins, J.H. Flower Garden Primer. NY, 1936. 1st ed. [x], (4) 333pp. VG in chppd dj. **$18.50\£11.00** *(RS 2/89)*

Cummins, Julia H. Flower Garden Primer, or Gardening Without Ears. NY, 1936. 333pp. Internally VG in slightly spttd clth. **$18.00\£10.00** *(BF 2/90)*

Cummins, Julia H. My Garden Comes of Age. NY, 1926. 180pp. Mild fxng else VG. **$20.00\£11.00** *(BF 2/90)*

Cunningham, G. Forest Flora of Canada. Canada, 1958. 144pp. Wrs. VG. **$10.00\£6.00** *(GW 8/89)*

Curtis, C.H. Orchids. London, 1950. [xi], 274pp. Bkplte, VG. **$60.00\£33.00** *(RS 2/89)*

Curtis, C.H. Orchids. London, 1950. 1st ed. 274pp. G. **$72.00\£40.00** *(ABx 3/89)*
Another copy. London, 1950. 1st ed. 274pp. Green cloth. Tips, head/foot of spine lightly worn else VG in yellowing dj. Badly chipped at head/foot. **$55.00\£31.00** *(CB 12/89)*

Curtis, W. Curtis's Botanical Magazine. Vol. XLII of 3rd Series, vol 12 of whole. 1886. 1/2 natural clf. Nice, clean cpy. **$810.00\£450.00** *(DL 89/90)*

Curtis, W. Practical Observations on the British Grasses. . . Likewise an Enumeration. London, 1812. 5th ed. 116pp. Orig brds, lack spine. Contents exc. **$110.00\£61.00** *(PBx 11/89)*

Curtis, William. Practical Observations on the British Grasses. London, 1812. 1790. 5th ed. 116pp. Publisher's boards with new spine. G. **$117.00\£65.00** *(ABx 3/89)*

Cushing, John. The Exotic Gardener. London, 1814. 3rd ed. [xii], 236pp. Mdrn paper rebnd, browning throughout. **$110.00\£61.00** *(Hrt 12/89)*

Cushman, J.A. Foraminifera of the Philippine and Adjacent Seas. Smithsonian, 1921. 608pp. Wraps. Few minor adhesions, some tears to wrps but never opened. **$22.00\£12.00** *(PBx 11/89)*

Cuthbert, Mabel. The Fall Flowers. Iowa, Brown, 1948. 199pp. Stiff wraps. Light wear. VG. **$8.00\£4.00** *(AB 4/89)*

D

"D.M." (Thomas Hyll). The Gardeners Labyrinth. Milford, 1939. 96pp. VG. **$40.00\£22.00** *(CB 12/89)*

Dahlgren, B.E. A Fossil Flower. Chicago, 1924. 16pp. Wrps. VG. **$15.00\£8.00** *(Hrt 12/89)*

Dale, I.R. & P.J. Greenway. Kenya Trees and Shrubs. Nairobi, 1961. [xxvii], 654pp. Black cloth, nick to back upper edge else VG. **$75.00\£42.00** *(BB 12/88)*

Dalla Torre, C.G. & H. Harms. Genera Siphonogamarum. 1900-07 (1963). 637pp. Folio. Reprnt without register. G. **$95.00\£53.00** *(PBx 11/89)*

Dallas, Ellen M. & Caroline A. Burgin. Among the Mushrooms. Toronto, 1900. 1st ed. 176pp. Very nice copy. Scarce. **$55.00\£31.00** *(JI 12/88)*

Dallimore, W. Holly, Yew and Box. Bodley Head, 1908. 284pp. Decorated cloth. G. **$81.00\£45.00** *(ABx 3/89)*

Dalziel, J.M. The Useful Plants of West Tropical Africa. London, 1955. [xii], 612pp. Ry 8vo. Wrpprs. G. **$54.00\£30.00** *(WW 1/90)*

Damon, Bertha. A Sense of Humus. NY, 1943. 250pp. VG in slightly chppd dj. **$15.00\£8.00** *(BF 2/90)*

Dana, Mrs. William Starr. How to Know the Wildflowers. NY, 1893. 1st ed. [xv], 298pp. VG. **$15.00\£8.00** *(RW 6/89)*
Another copy. NY, 1906. [xxxix], 346pp. VG. **$8.00\£4.00** *(CB 12/89)*

Dana, Mrs. William Starr. Plants and Their Children. NY, 1896. 272pp. VG. **$10.00\£6.00** *(RW 6/89)*

Dandy, J.E. List of British Vascular Plants, Incorporating the London Catalogue. London, 1958. Interleaved. G. **$21.00\£12.00** *(WW 1/90)*

Daniel, L. Études sur la Greffe. 2 text vols & 1 plt vol., 1927-30. Total 1,377 pp. Very scarce. Only 100 sets pblshd. Nmbrd, sgnd by author. Hardly used. **$175.00\£97.00** *(PBx 11/89)*

Daniels, Francis Potter. The Flora of Boulder, Colorado and Vicinity. Columbia, MO, 1911. [xiv], 311pp. Wrapper chipped lower front corner.

Ink names on wrapper. Scarce. **$25.00\£14.00** *(BB 12/88)*

Dannenfeldt, K. Leonhard Rauwolf: 16th Century Physician, Botanist & Traveler. Harvard, 1968. 321pp. VG in dj. **$12.50\£7.00** *(GW 8/89)*

Dannenfeldt, Karl H. Leonard Rauwolf, Sixteenth-Century Physician, Botanist and Traveler. Cambridge, Mass, 1968. [viii], 321pp. F in dusty dj. **$24.00\£13.00** *(Hrt 12/89)*

Darnell, A.W. Winter Blossoms from the Outdoor Garden. London, 1926. 335pp. G. **$27.00\£15.00** *(ABx 3/89)*

Darrow, George. The Strawberry. NY. Holt, 1966. 447pp. As New in worn dj. **$30.00\£17.00** *(AB 2/89)*
Another copy. NY, 1966. [xvi] + 447pp. VG in dj. **$27.50\£15.00** *(WH 1/90)*

Darwin, Charles. Insectivorous Plants. 1893. 2nd ed. Amer ed. VG. **$54.00\£30.00** *(DL 89/90)*

Darwin, Erasmus. Phytologia. London, 1800. [viii], 612pp + index. Some fxing first and last few pp. New 1/2 leather/brds. **$220.00\£122.00** *(Hrt 12/89)*

Davidson, John. Conifers, Junipers and Yew. London, 1927. 1st ed. [xviii], 74pp. VG. **$55.00\£31.00** *(JI 12/88)*

Davie, W. Galsworthy & W. Curtis Green. Old Cottages and Farmhouses in Surrey. London, 1908. Small 4to, gilt decorated green cloth. 69pp. Spine faded, moderate wear. **$75.00\£42.00** *(MD 11/88)*

Davis, Clare O. In Our Country Garden. NY, 1938. 1st ed. 207pp. Decorated endpapers. VG in dj. **$5.00\£3.00** *(RW 11/86)*

Davis, H. Life & Work of Cyrus Guernsey Pringle (Botanist). VT, 1936. 756pp. Wrs. VG. **$16.00\£9.00** *(GW 8/89)*

Davis, Lucius. Ornamental Shrubs. NY, 1899. [vi], 338pp. G. **$72.00\£40.00** *(ABx 3/89)*
Another copy. NY, 1899. Small 4to, gilt decor. blue cloth. 338pp. T.e.g., nice copy with a few minor spots on cover. **$22.50\£13.00** *(MD 11/88)*
Another copy. NY, 1899. 1st ed. [viii], 342pp. T.e.g., sm 4to. Sl shelf wear else F. **$115.00\£64.00** *(JI 12/88)*

Davis, P.H. Flora of Turkey and the East Aegean Islands. Vol. I, 1956, 1965. VG. **$121.00\£67.00** *(DL 89/90)*

Davison, Ralph. Concrete Pottery and Garden Furniture. NY, 1910. [xiii], 196pp advts. Some staining to brds, note pasted to inside frnt brd. **$60.00\£33.00** *(Hrt 12/89)*

Dawson, E. Marine Botany: An Introduction. Holt, 1966. 371pp. VG. **$18.00\£10.00** *(GW 8/89)*

De Crespigny, E.C. A New London Flora. London, 1877. [xxiv], 179pp. Clth. G. **$27.00\£15.00** *(WW 1/90)*

De Graff, Jan & Edward Hyams. Lilies. Np, 1968. 142pp. Gilt embossed red cloth. Fine in dj. **$16.00\£9.00** *(BB 12/88)*

De Lille, Jacques (l'Abbe De Lille). Les Jardins, ou L'Art d'Embellir les Paysages. Paris and Reims: Cazin, confusingly marked as 5th ed (1791) and 6th ed (1785). Engraved frontispiece and vignette on the Parisian titlepage. [x], 138, (1) pp. Internally VG in quite wrn but solid speckled clf. **$75.00\£42.00** *(BF 2/90)*

De Mauny the Count. The Gardens of Taprobane. 1st and only ed. 1937. Orig clth. Scarce. VG. **$36.00\£20.00** *(DL 89/90)*

De Sounin, Leonie. Magic in Herbs. NY, 1952. 208pp. VG in dj. **$5.00\£3.00** *(RW 11/86)*

Deakin, R. The Ferns of Britain. London, 1848. 138pp. Brds. G. **$32.00\£18.00** *(WW 1/90)*

Deakin, Richard. Flora of the Colosseum of Rome. London, 1873. [viii], 237pp. A.e.g. G. **$81.00\£45.00** *(ABx 3/89)*

Deakin, Richard. Florigraphia Britannica: or Engravings and Descriptions of the Flowering Plants and Ferns of Britain. 1857. 4 vols. Orig clth. Nice clean set. **$630.00\£350.00** *(DL 89/90)*

Deam, Charles C. Flora of Indiana. Indianapolis, 1940. 1236pp. Blue cloth, larger format (7 1/4 x 10 1/4) than reprints. Near fine. **$68.00\£38.00** *(BB 12/88)*

Dean, A. Root and Stem Vegetables. Nd, circa 1912. Scarce. F. **$16.00\£8.50 DL 89/90)**

Deckert, Helmut. Maria Sibylla Merian's "Neues Blumenbuch". (Nurnberg, 1680). Leipzig, 1966. 2 vols. in slip case. Vol I: 72pp of text in German. Vol II: 36 colour plates. G. **$94.00\£52.00** *(ABx 3/89)*

Deerr, Noel. Cane Sugar. Altringham, Manchester, 1911. [xv], 592pp. Glt stmpd brds, sl tender, handled. Else VG. **$45.00\£25.00** *(Hrt 12/89)*

Delitzsch, Franz. Iris: Studies in Colour and Talks about Flowers. Edinburgh, 1889. 227pp. Orig bndng. VG. **$24.00\£13.00** *(Hrt 12/89)*

Delorit, Richard J. An Illustrated Taxonomy Manual of Weed Seeds. River Falls, WI, 1970. 175pp. Plastic spiral binding. Ex-ref. lib. **$14.00\£8.00** *(BB 12/88)*

Dennis, R.W.G. British Cup Fungi and Their Allies. London, 1960. xxiv, 280pp. VG in dj. **$45.00\£25.00** *(Hrt 12/89)*

Dies, Edward Jerome. Titans of the Soil. U of S. Carolina Press, 1949. 213pp. VG in worn dj. **$11.00\£6.00** *(CB 12/89)*

Dietz, Marjorie. Landscaping and the Small Garden. Garden City, 1973. 170pp. G in G dj. **$9.00\£5.00** *(CB 12/89)*

Dietz Press (publishers). Photographic Studies of Old Virginia Homes and Gardens. London, nd (1942?). 49 photos. G. **$27.00\£15.00** *(ABx 3/89)*

Digilio, A.P.L. & E.P. Legname. Los Arboles Indigenas de la Provincia de Tucuman. Opera Lilloana XV, 1966. 278pp. Sewed. G. **$56.00\£31.00** *(WW 1/90)*

Dillistone, G. (ed). Dykes on Irises. London, 1930. [ix], 294pp. Annotations on rear endpapers. G. **$36.00\£20.00** *(WW 1/90)*

Dillon, Julia. Blossom Circle of the Year in Southern Gardens. NY, 1922. 202pp. As New. **$18.00\£10.00** *(AB 2/89)*

Dinges & Conard. The New Guide to Rose Culture. West Grove, PA, 1879. 55pp. Wraps and contents nearly new. Scarce. **$55.00\£31.00** *(AB 4/89)*

(Dioscorides) Berendes, J. Des Pedanios Dioskurides aus Anazarbos Arzneimittellehre in Fuenf Buechern. Stuttgart, 1902. 572pp. G. **$185.00\£103.00** *(PBx 11/89)*

Ditchfield, P.H. The Manor Houses of England. NY, 1910. Small 4to, rubbed blue cloth. 211pp. T.e.g., shelf worn, somewhat shaken. **$30.00\£17.00** *(MD 11/88)*

Dixon, H.N. The Student Handbook of British Mosses. London, 1954. Reprt of 3rd ed. 582pp. Brds stained & very mild interior fxng. **$24.00\£13.00** *(Hrt 12/89)*

Dobyns, Winifred Starr. California Gardens. NY, 1931. 20pp. of text. 208 b&w plts. Internally VG in soiled brds, spine ends frayed. Sgnd by Dobyns on e.p. **$40.00\£22.00** *(BF 2/90)*

Dodge, C.W. Some Lichens of Tropical Africa V. London, 1971. [ii], 226pp. Wrpprs. G. **$22.00\£12.00** *(WW 1/90)*

Donk, M.A. Revision der Niederlandischen Heterobasidiomycetae und Homobasidiomycetaeaphyllophoraceae. Lehre, 1969. 416pp. Wrpprs. G. **$32.00\£18.00** *(WW 1/90)*

Dore, W.G. Grasses of the Ottawa District. 1959. Pbnd. G. **$8.00\£4.00** *(PBx 11/89)*

Dorman, Caroline. Flowers Native to the Deep South. Baton Rouge, 1959. [xvii], 176pp. VG in dj. **$14.00\£8.00** *(BB 12/88)*

Dorrance, Ann. Fragrance in the Garden. NY, 1937. 96pp. VG in sunned dj. **$12.00\£7.00** *(AB 4/89)*

Dorrance, Anne. Green Cargoes. NY, 1945. 187pp. VG. **$10.00\£6.00** *(RW 6/89)*

Dougherty, Betty. Green Gardener. London, 1975. 200pp. G in dj. **$11.00\£6.00** *(ABx 3/89)*

Douglas, David. Journal Kept by David Douglas During His Travels in North America 1823-1827. NY. Antiquarian Press, 1959. Limited to 750 copies. 376pp. T.e.g. VG. **$65.00\£36.00** *(JI 12/88)*

Douglas, Marjory. Adventures in a Green World. Miami, 1973. 61pp. VG in dj. **$20.00\£11.00** *(AB 4/89)*

Downing, A.J. A Treatise on the Theory and Practice of Landscape Gardening. NY, 1967. Facsimile edition. [xvii], 576pp. VG in dj. **$25.00\£14.00** *(RW 11/86)*

Drewitt, F. Dawtrey. The Romance of the Apothecaries' Garden at Chelsea. London, 1924. 2nd ed. Rev. and enlgd. 136pp. Scarce. **$30.00\£17.00** *(RW 11/86)*

Another copy. London, 1924. 2nd ed. 136pp. G. **$29.00\£16.00** *(ABx 3/89)*

Dreyer, Peter. A Gardener Touched with Genius. NY, 1975. 322pp. G in dj. **$32.00\£18.00** *(ABx 3/89)*

Druce, G.C. The Flora of Buckinghamshire. Arbroath, 1926. [cxxviii], 437pp. Ry 8vo. Cloth. Cheap cpy, lacking map. **$45.00\£25.00** *(WW 1/90)*

Druce, G.C. The Flora of Oxfordshire. Oxford, 1927. 2nd ed. [cxxxii], 538pp. Clth, trifle used. **$144.00\£80.00** *(WW 1/90)*

Du Mont de Courset, G.L.M. Le Botaniste Cultivateur. 2e Edition, Considérablement Augmentée. Paris, 1811. 6 vols. Hlf clf. Nice copy. **$153.00\£85.00** *(WW 1/90)*

DuCane, Florence. Flowers & Gardens of Madeira. London, 1926. 150pp. Nearly New. **$25.00\£14.00** *(AB 2/89)*

Duddington, C. Beginners Guide to Botany. Drake, 1970. 200pp. VG in dj. **$8.75\£5.00** *(GW 12/88)*

Duggar, B.M. Mushroom Growing. NY, 1925. 250pp. Cloth light wear else VG. **$12.00\£7.00** *(AB 4/89)*

Duncan, F.M. Plants & Their Children. London, 1936. 78pp. Spine faded else VG. **$10.00\£6.00** *(AB 2/89)*

Duncan, Frances. The Joyous Art of Gardening. NY, 1917. [xiii], 239pp. Dec cvr. VG. **$10.00\£6.00** *(RW 11/86)*

Duncan, U.K. A Guide to the Study of Lichens. Arbroath, 1959-63. [xxvii], 164pp. Lichen illustrations supplement, 142 plates. Together, 2 vols. G. **$54.00\£30.00** *(WW 1/90)*

Dunham, Elizabeth. How to Know the Mosses. Boston, 1916. [xxv], 287pp. G. **$15.00\£8.00** *(WH 1/90)*

Dunsterville, G. Introduction to the World of Orchids. Doubleday, 1964. 102pp. VG. **$18.00\£10.00** *(GW 12/88)*

Dunsterville, G.C.K. & L. Garay. Venezuelan Orchids Illustrated. London, 1959. Vol. 1. 452pp. G. **$108.00\£60.00** *(WW 1/90)*

Dunthorne, G. Flower and Fruit Prints of the 18th and Early 19th Centuries, Their Histories, Makers and Uses, with a Catalogue Raisonne of the Works in which They Are Found. (1938) facs ed. 1970. Sm folio. VG in buckrm. **$216.00\£120.00** *(DL 89/90)*

Durand, Herb. My Wild Flower Garden. NY, 1927. 242pp. Cover spotted, signed w/inscrp by author. **$15.00\£8.00** *(AB 2/89)*

Durand, T. Index Generum Phanerogamarum. Brussels, 1888. [xxii], 722pp. [l]. New clth, stmp on title page. **$63.00\£35.00** *(WW 1/90)*

Dutton, Joan. The Flower World of Williamsburg. NY, 1962. 148pp. VG. **$12.00\£7.00** *(AB 4/89)*

Dutton, R. The English Garden. London, 1950. 2nd ed. G. **$27.00\£15.00** *(WW 1/90)*
Another copy. London, 1950. 2nd ed. 122pp. VG in remains of dj. **$25.00\£14.00** *(BF 2/90)*

Dutton, Ralph. The English Garden. London, 1950. 2nd rev. ed. [vi], 122pp. VG in dj. **$20.00\£11.00** *(RW 11/86)*
Another copy. London, 1950. 2nd ed. 122pp. VG in dj. **$15.00\£8.00** *(AB 4/89)*
Another copy. London, 1945. 2nd ed. 122pp. G in torn dj. **$18.00\£10.00** *(ABx 3/89)*

Dykes, W.R. A Handbook of Garden Irises. London, 1924. 250pp. Spine faded else G. **$36.00\£20.00** *(ABx 3/89)*

Dykes, W.R. Notes on Tulip Species. London, 1930. 108pp. Head and tail of spine expertly repaired, dj worn and repaired. Internally very clean. Folio. **$252.00\£140.00** *(ABx 3/89)*

E

Earle, Alice Morse. Home Life in Colonial Days. NY, 1900. 5th prntng. 8vo. [xvi], 470pp. Light wear else VG. **$12.00\£7.00** *(CB 12/89)*

Earle, Alice Morse. Old Time Gardens. NY, 1901. 1st ed. [xviii], 489pp. G. **$15.00\£8.00** *(CB 12/89)*

Earle, Alice Morse. Old Time Gardens. NY, 1901. 489pp. VG in slightly rbbd pict clth. **$35.00\£19.00** *(BF 2/90)*

Earle, Alice Morse. Sun-Dials and Roses of Yesterday. NY, 1902. [xxiii], 461pp. Decorated cover. T.e.g. VG. **$35.00\£19.00** *(RW 11/86)*
Another copy. NY, 1902. 1st ed. 461pp. Decorated cloth. As New. **$50.00\£28.00** *(AB 4/89)*
Another copy. NY, 1902. 1st ed. [xx], 462pp. VG. **$60.00\£33.00** *(JI 7/88)*

Earle, Alice Morse. Sundials and Roses of Yesterday. NY, 1922. 461pp. Bndng slightly rbbd else VG. **$30.00\£17.00** *(BF 2/90)*

Earle, Mrs. C.W. & Miss Ethel Case. Pot-Pourri Mixed by Two. London, 1914. 1st ed. [xiii], 456pp. G. **$22.00\£12.00** *(CB 12/89)*

Earle, Mrs. C.W. & Miss Ethel Case. Potpourri Mixed by Two. NY, 1914. 1st Amer ed. 456pp. Bndng slightly rbbd, spine faded. **$30.00\£17.00** *(BF 2/90)*

Earle, Mrs. C.W. A Third Pot-pourri. London, 1903. 2nd ed. 440pp. VG. **$30.00\£17.00** *(BF 2/90)*

Earle, Mrs. C.W., E.V.B. et al. Garden Colour. NY, 1922. [xvi], 192pp. VG. **$45.00\£25.00** *(RW 6/89)*

Earle, Mrs. C.W. More Pot-pourri from a Surrey Garden. NY, 1899. 1st Amer ed. 463pp. Ownr's sig else VG in faded clth, spine spttd. **$30.00\£17.00** *(BF 2/90)*

Earle, Mrs. C.W. More Pot-Pourri from a Surrey Garden. London, 1899. 2nd imp. 453pp. Spine faded else G. **$27.00\£15.00** *(ABx 3/89)*
Another copy. London, 1899. 1st ed. 453pp + ads. G only. **$20.00\£11.00** *(CB 12/89)*

Earle, Mrs. C.W. Pot-Pourri from a Surrey Garden. London, 1897. 3rd ed. 381pp. Spine faded and brittle at head. G. **$22.00\£12.00** *(ABx 3/89)*
Another copy. London, 1912. 13th ed. [xii], 381pp. Near F. **$30.00\£17.00** *(Hrt 12/89)*

Earle, Mrs. C.W. Potpourri from a Surrey Garden. London, 1899. 20th ed. 381pp. T.e.g., green clth slightly spttd. **$35.00\£19.00** *(BF 2/90)*

Eastwood, Alice. A Collection of Popular Articles on the Flora of Mount Tamalpais. Np, 1944. 32pp. Softbound, tan leatherette. Fine. **$16.00\£9.00** *(BB 12/88)*

Eastwood, Dorothea. Mirror of Flowers. Boston, 1953. 1st Amer ed. 237pp. VG. **$20.00\£11.00** *(BF 2/90)*

Eastwood, Dorothea. The Story of Our Gardens. Bedford, 1958. 12mo. [vii], 92pp. As New in dj. **$10.00\£6.00** *(RW 11/86)*

Eaton, Amos M. & J. Wright. North American Botany. . . Native & Common Plants North of Mexico. Troy, NY, 1840. 8th ed. 625pp. Fll clf. Nice and sound. **$62.00\£34.00** *(PBx 11/89)*

Eaton, Walter P. Everybody's Garden. NY, 1932. 1st ed. 169pp. VG. **$15.00\£8.00** *(RW 6/89)*

Eaton, Walter P. Wild Gardens of New England. Boston, 1936. 12mo. 124pp. VG in worn dj. **$15.00\£8.00** *(RW 6/89)*

Eden, F. A Garden in Venice. London, 1903. 135pp. Vellum spine, paper boards. **$104.50\£58.00** *(ABx 3/89)*

Eden, H.K.F. & Eleanor Lloyd (eds). The Book of Sun-Dials. London, 1900. 4th ed. 529pp. Gilt lettering dull else VG. Bookplate of Lord Elphinstone George Bell. **$270.00\£150.00** *(ABx 3/89)*

Edgarton, Miss S.C. The Flower Vase; Containing the Language of Flowers, and Their Poetic Sentiments. Lowell, MA, 1844. 157pp. Bkplt, edge wrn else near VG. **$35.00\£19.00** *(BF 2/90)*

Edwards, A. & D. Bogdan. Important Grassland Plants of Kenya. Pitman, 1951. 124pp. VG. **$16.00\£9.00** *(GW 12/88)*

Edwards, A. Rock Gardens. London, nd (c1929). 320pp. 8 tipped-in color plates, 50 photos and diagrams. VG. **$15.00\£8.00** *(RW 11/86)*

Edwards, Paul. English Garden Ornament. Barnes, S. Brunswick & NY, 1965. 110pp. F in reprd dj. **$18.00\£10.00** *(CB 1/90)*

Edwards, Paul. English Garden Ornament. S. Brunswick, 1965. 110pp. Green cloth, VG in wrn DJ. **$22.50\£13.00** *(MD 2/90)*

Eggers, H. The Flora of St. Croix and the Virgin Islands. USNM, 1879. 133pp. Bound. Library stamps else VG. **$20.00\£11.00** *(GW 12/88)*

Eifert, Virginia. Tall Trees and Far Horizons. NY, 1965. [xvii], 301pp. VG in dj. Ann Leighton's bkplte. **$20.00\£11.00** *(RW 6/89)*

Elder, Walter. The Cottage Garden of America. Phila., 1850. 2nd ed., rev. 233pp. Light wear, some marginal water stains. Scarce. **$45.00\£25.00** *(AB 4/89)*

Elgood, G.S. & G. Jekyll. Some English Gardens. 1904. 1st ed. Folio. Orig clth, t.e.g. Scarce. VG. **$270.00\£150.00** *(DL 89/90)*

Elgood, George S. Italian Gardens, After Drawings by George S. Elgood, R.I., with Notes by the Artist. London, 1907. 157pp. Folio. Scttrd fxng, quite flaked else VG. **$125.00\£69.00** *(BF 2/90)*

Eliovson, Sima. Discovering Wild Flowers in Southern Africa. Cape Town, 1962. 253pp. 4to, white cloth, end paper maps. Slight edge wear to dj, else VG. **$43.50\£24.00** *(BB 12/88)*

Eliovson, Sima. Flowering Shrubs and Trees for Gardens in Southern Africa. 1952. 1st ed. **$45.00\£25.00** *(DL 89/90)*

Eliovson, Sima. South African Wild Flowers for the Garden. Cape Town, 1965. 4th ed. (1955). [xix], 306pp. 4to. Front cover crease, else VG in dj. **$38.00\£21.00** *(BB 12/88)*
Another copy. Timmins, 1957. 306pp. 2nd ed. VG. **$50.00\£28.00** *(GW 8/89)*

Elliott, F.R. Elliott's Fruit Book. NY, 1856 (1854). 503pp. Cloth worn, owner's name on title page, internally As New. **$45.00\£25.00** *(AB 2/89)*

Elliott, J. Wilkinson. A Plea for Hardy Plants. NY, 1907. 96pp. Cloth & contents As New. **$20.00\£11.00** *(AB 2/89)*
Another copy. NY, 1902. Small 4to. 76pp. Dec cover. VG. **$20.00\£11.00** *(RW 11/86)*

Elliott, J. Wilkinson. Adventures of a Horticulturist. Point Loma, CA. Privately printed, 1935. 1st ed. Signed by author. 232pp. Slight cover wear and foxing else VG. **$45.00\£25.00** *(JI 7/88)*
Another copy. Point Loma, CA. By the author, 1935. 232pp. Signed. As New. **$10.00\£6.00** *(AB 2/89)*

Elliott, Simon B. The Important Timber Trees of the U.S. Boston, 1912. [xix], 382pp. Gold embossed green cloth. G. **$16.00\£9.00** *(BB 12/88)*

Ellis, Aytoun. The Essence of Beauty, A History of Perfume & Cosmetics. London, 1960. [xvii], 200pp. Presentation note on f.f.e.p.s else VG in trn dj. **$30.00\£17.00** *(Hrt 12/89)*

Ellwanger, H.B. The Rose. Rev. ed. 1901. Brds. 310pp. VG. **$35.00\£19.00** *(PBx 11/89)*

Ely, Helena R. Another Hardy Garden Book. NY, 1905. 1st ed. [xv], 243pp, [ii]. Spine faded else VG. **$10.00\£6.00** *(RW 6/89)*

Ely, Helena R. The Practical Flower Garden. NY, 1911. Reprint. 304pp. G. **$21.50\£12.00** *(ABx 3/89)*

Emerson, Edward. The Story of the Vine. NY, 1902. 252pp. VG. **$7.50\£4.00** *(WH 1/90)*

Emerson, George & C. Flint. Manual of Agriculture. Boston, 1862. 306pp. Cloth, contents uniformly waterstained else VG. **$18.00\£10.00** *(AB 2/89)*

Engel, David H. Japanese Gardens for Today. Rutland, 1959. 270pp. VG in pict clth. **$50.00\£28.00** *(BF 2/90)*
Another copy. Rutland, VT, 1959. [xvi], 250pp. Rebckd, neat ex-libry. Dec clth. **$30.00\£17.00** *(CB 1/90)*

Engel, David. Japanese Gardens for Today. Rutland, 1959. Dec brown cloth. 270pp. F in dj and original box. **$45.00\£25.00** *(MD 11/88)*
Another copy. Rutland, 1959. [xvi], 250pp. 4to. Re-backed, neat ex-library. **$30.00\£17.00** *(CB 12/89)*

Engler, A. and K. Prantl. Die Naturlichen Pflanzenfamilien. Vol. 13, Gymnospermae. Berlin, 1926. Reprint 1960. 2nd ed. [iv], 447pp. Clean 2nd hand cpy. **$63.00\£35.00** *(WW 1/90)*

Engler, A. Monographie der Gattung Saxifraga. L. Breslau, 1872. 292pp. Hlf clth (library bookplate). G. **$45.00\£25.00** *(WW 1/90)*

English Forests and Forest Trees. London, 1853. [xii], 406pp. Good. **$81.00\£45.00** *(ABx 3/89)*

Erickson, R. Orchids of the West. Perth, 1951. 109pp. G. **$32.00\£18.00** *(WW 1/90)*

Esau, K. Plant Anatomy. NY, 1965. 2nd ed. 767pp. F in dj. **$15.00\£8.00** *(WH 1/90)*

Eschnauer, H. Spurenelemente in Wein & Anderen Getraenken. 1974. Np. G in dj. **$22.00\£12.00** *(PBx 11/89)*

Esper, E.J. Icones Fucorum. London, 1966. 358pp. G. **$72.00\£40.00** *(WW 1/90)*

Estienne, Ch. De Re Hortensi Libellus. Lyon, 1539. 12mo. Modern wraps. 97, [15]pp. Crisp, clean. **$575.00\£319.00** *(PBx 11/89)*

Ettl, H. Die Gattung Chloromonas Gobi Emend. Wille. Beih. Nov. Hedw. 34, 1970. [iv], [xvi], 284pp. Wrpprs. G. **$27.00\£15.00** *(WW 1/90)*

Evelyn, John. Kalendarium Hortense, or, The Gard'ner's Almanac. Falls Village, CT. #171 of 485 cpies. Facs. Brwn clth. Inscribed by the printer to artist Armin Landeck. F. **$35.00\£19.00** *(BF 2/90)*

Evelyn, John. Memoires for My Grandson. London, Nonesuch Press, 1926. Ltd. ed. of 1250. #597. Handmade paper, VG in mrbled aging tape reprd box. **$85.00\£47.00** *(Hrt 12/89)*

Evelyn, John. Sylva, or A Discourse on Trees. London, 1670. 2nd ed. [xlviii], 247 (recte 241) pp, (1); [iv], 67pp; 33 (1). Contemp lthr. Very nice cpy. **$575.00\£319.00** *(Hrt 12/89)*
Another copy. London. 1706. 4th ed. pp. [xxv], 384, [iii], 275, [i]. Contemp panelled clf, trifle wrn. Neatly rebckd. Peter Collinson's cpy. Inscrip by Dawson Turner. Drwng by Collinson inserted and 2 engrvngs. **$756.00\£420.00** *(WW 1/90)*

Evelyn, John. Sylva. London, 1670. 2nd ed. [xlvii], 247, 67, 33pp (separate title pages). Fine old speckled calf; base of spine worn. Last 2 pages repaired, slight loss of errata. **$342.00\£190.00** *(ABx 3/89)*

Everard, B. Wild Flowers of the World. London, 1974. New ed. G. **$30.00\£17.00** *(WW 1/90)*

Everett, Katherine. Bricks and Flowers. London, 1950. [x], 252pp. VG. **$10.00\£6.00** *(RW 6/89)*

Everett, T. Living Trees of the World. Doubleday, 1968. 315pp. VG in dj. **$15.00\£8.00** *(GW 8/89)*

Ewan, Joseph and Nesta. John Banister, and His Natural History of Virginia 1678-1692. Urbana, 1970. [xxx], 485pp. F in dj. **$45.00\£25.00** *(Hrt 12/89)*

Ewan, Joseph and Nesta. John Bannister and His Natural History of Virginia 1678-1692. Urbana, 1970. 1st ed. [xxx], 490pp. As New in dj. **$35.00\£19.00** *(JI 7/88)*

Ewart, Alfred J. Handbook of Forest Trees for Victorian Foresters. Melbourne, c1925. 1st ed. 524pp. Some fading, back hinge starting. **$35.00\£19.00** *(JI 12/88)*

Excell, A.W. Catalogue of the Vascular Plants of St. Tomé, with Principe & Annobon. BM/NH, 1944. 428pp. F. **$42.00\£23.00** *(PBx 11/89)*

Eyre, S.R. Vegetation and Soils. Chicago, 1968. 2nd ed. 328pp. F in dj. **$10.00\£6.00** *(WH 1/90)*

F

Faegri, K. and J. Iversen. Textbook of Pollen Analysis. London, 1964. 2nd ed. G. **$21.00\£12.00** *(WW 1/90)*

Fairbridge, D. Gardens of South Africa. 1924. 1st ed. Orig clth. G. **$32.00\£17.50 DL 89/90)**

Fairbrother, Nan. An English Year. NY, 1954. 243pp. VG in dj. **$20.00\£11.00** *(RW 6/89)*

Fairbrother, Nan. Men and Gardens. London, 1956. 1st ed. 240pp. G in dj. **$29.00\£16.00** *(ABx 3/89)*
Another copy. NY, 1956. [xii], 271pp. VG in dj. **$25.00\£14.00** *(RW 6/89)*
Another copy. NY, 1956. 1st Amer. [xii], 271pp. Ink sig else VG in torn dj. **$40.00\£22.00** *(Hrt 12/89)*

Fairbrother, Nan. Men and Gardens. NY, 1956. 1st Amer. [xii], 271pp. Ink sig else VG in dj. **$40.00\£22.00** *(Hrt 12/89)*
Another copy. London, 1956. 240pp. F in slightly wrn dj. **$35.00\£19.00** *(BF 2/90)*

Fairbrother, Nan. The House in the Country. NY, 1963. 1st Am. ed. 237pp. VG in dj. **$25.00\£14.00** *(RW 6/89)*

Fairchild, David. Garden Islands of the Great East. NY, 1943. [xiv], 239pp. Nice, brght cpy. **$27.00\£15.00** *(Hrt 12/89)*
Another copy. NY, 1943. 1st ed. [xiv], 239pp. Few very lght spts to front covr else VG. **$15.00\£8.00** *(CB 1/90)*

Fairchild, David. Garden Islands of the Great East. NY, 1945. [xiv], 239pp. VG in dj. **$10.00\£6.00** *(RW 11/86)*
Another copy. NY, 1943. 1st ed. 239pp. VG. **$15.00\£8.00** *(CB 12/89)*
Another copy. NY, 1944. [xvi], 240pp. Cover worn else VG. **$15.00\£8.00** *(JI 12/88)*

Fairchild, David. The World Grows Round My Door. NY, 1947. 1st ed. [xii], 347pp. Ownr's name f.f.e.p. Lghtly soiled, a few spts. G. **$30.00\£17.00** *(CB 1/90)*

Fairchild, David. The World Grows Round My Door. NY, 1947. 1st ed. 347pp. G in repaired dj. **$43.00\£24.00** *(ABx 3/89)*
Another copy. NY, 1947. [xii], 347pp. G. **$30.00\£16.00** *(CB 8/89)*

Fairchild, David. The World Was My Garden. NY, 1938. 1st ed. [xiv], 494pp. VG. **$15.00\£8.00** *(RW 6/89)*

Fairchild, David. The World Was My Garden. NY, 1939. 494pp. Lght wr. VG. **$20.00\£11.00** *(CB 1/90)*

Falkner, F. Farmer's Manual; Practical Treatise on the Nature and Value of Manures. NY, 1843. 153pp. Cloth lightly spotted, contents with water stain at lower margin. **$20.00\£11.00** *(AB 2/89)*

Faris, John T. Old Gardens in and About Philadelphia. Indianapolis, 1932. 311pp. Tips bmpd else VG. **$35.00\£19.00** *(CB 1/90)*

Faris, John. Old Gardens In and About Philadelphia. Indianapolis, 1932. 311pp. VG. **$20.00\£11.00** *(RW 11/86)*
Another copy. Indianopolis, 1932. 311pp. Rebound in dark green quarter calf and buckram. VG. **$72.00\£40.00** *(ABx 3/89)*

Farrer, R. Among the Hills. London, 1911. 326pp. Orig clth, slightly used. Scarce orig issue. Lght fxing, names on e.p. **$108.00\£60.00** *(WW 1/90)*

Farrer, R. My Rock Garden. London, 1913. 5th imp. Orig blck clth G. **$32.00\£18.00** *(WW 1/90)*

Farrer, R. The Dolomites: King Laurin's Garden. London, 1913. [viii], 207pp. Orig dec clth. G. **$81.00\£45.00** *(WW 1/90)*

Farrer, R. The Rainbow Bridge. London, 1921. [xi], 383pp. Orig clth, trifle used. 1st imp. Lght fxing. **$108.00\£60.00** *(WW 1/90)*
Another copy. London. 1921. 383pp. Newly bnd, glt stmpd hlf lthr. Some fxng first few pp frnt & bck else VG. **$105.00\£58.00** *(Hrt 12/89)*

Farrer, Reginald. On the Eaves of the World. London, 1917. 1st ed. 2 vols. [xii], 311pp; 328pp. Both vols slightly worn and dull. **$153.00\£85.00** *(ABx 3/89)*

Farrer, Reginald. The Dolomites – King Laurin's Garden. 1913. 1st & only ed. Becoming scarce. Well preservd cpy. **$81.00\£45.00** *(DL 89/90)*

Farrer, Reginald. The English Rock Garden. London, 1930. 2 vols. [lxiv], 504pp; [viii], 524pp. S.S. Berry's set. Minimal lower corner wear. Slightly chipped djs. **$115.00\£64.00** *(BB 12/88)*
Another copy. London, 1930. 5th imp. 2 vols. [lxiv], 504pp; 524pp. G. **$117.00\£65.00** *(ABx 3/89)*

Farrer, Reginald. The Plant Introductions of Reginald Farrer. London, 1930. [xi], 113pp. 4to, beveled white buckram, top edge gilt. Ltd. to 500 copies. 450 were for sale. Very scarce. Stamp of S.S. Berry on pastedown. VG. **$275.00\£153.00** *(BB 12/88)*

Farrer, Reginald. The Rainbow Bridge. London, 1921. 1st ed. 383pp. G. **$99.00\£55.00** *(ABx 3/89)*

Farrer, Reginald. The Rock Garden. London, nd (c1912). [xi], 118pp. Green cloth. G. **$20.00\£11.00** *(BB 12/88)*

Farrington, Edward (ed). The Gardener's Omnibus. Boston, 1938. 4to. [xviii], 886pp. VG in dj. **$20.00\£11.00** *(RW 11/86)*

Farrington, Edward. Ernest H. Wilson, Plant Hunter. Boston, 1931. [xxi], 197pp. Nr F in older dj. **$40.00\£22.00** *(Hrt 12/89)*

Farwell, Edith F. My Garden Gate Is on the Latch. Lake Forest, IL, 1962. 77pp. As New in dj. Signed. **$10.00\£6.00** *(AB 2/89)*

Farwell, Edith Foster. My Garden Gate Is on the Latch. Lake Forest, IL, 1962. Pblshd for the author. 77pp. F in dj. **$12.00\£7.00** *(BF 2/90)*

Fauconnet, G.P. Flower – Name Fancies. London & NY, 1918. 80pp. G. **$36.00\£20.00** *(ABx 3/89)*

Faure, Gabriel. Gardens of Rome. Fair Lawn, NJ, 1960. 1st thus. 201pp. Grey clth. VG in chippd, slightly trn dj. **$40.00\£22.00** *(CB 1/90)*

Faure, Gabriel. Gardens of Rome. London, 1960. 201pp. VG in dj. **$29.00\£16.00** *(ABx 3/89)*
Another copy. Fair Lawn, NJ, 1960. Oxford imprt. 1st thus. 201pp. VG in chipped, slightly torn dj. **$40.00\£22.00** *(CB 12/89)*

Favretti, Rudy F. & Gordon P. De Wolf. Colonial Gardens. Barre, MA, 1972. 163pp. Ed of 4000. VG in dj. **$20.00\£11.00** *(BF 2/90)*

Favretti, Rudy J. New England Colonial Gardens. Stonington, CT. Pequot Press Handbook #1. Wraps. 32pp. Edges browning else VG. **$10.00\£6.00** *(CB 12/89)*

Feininger, Andreas. Trees. NY, 1968. 160pp. of photos, line drawings, 7 maps. VG in dj. **$55.00\£31.00** *(RW 6/89)*

Feldmaier, Carl. Lilies. London, 1970. 1st Eng. ed. [x], 228pp. VG in tattered (piece torn from back) dj. Inscription f.f.e.p. from Park seed co. **$15.00\£8.00** *(CB 12/89)*

Felsko, E. A Book of Wild Flowers. Oxford, 1956. A little minor foxing. G. **$27.00\£15.00** *(WW 1/90)*

Felt, Ephraim & W. Howard Rankin. Insects and Diseases of Ornamental Trees and Shrubs. NY, 1932. [xix], 507pp. VG. **$17.50\£10.00** *(BB 12/88)*

Felt, Ephraim P. Our Shade Trees. NY, 1942. 2nd ed. 316pp. VG. **$15.00\£8.00** *(RW 6/89)*

Felton, R.F. British Floral Decoration. London, 1910. 194pp. G. **$68.50\£38.00** *(ABx 3/89)*

Fernov, Bernhard E. The Care of Trees in Lawn, Street and Park. NY, 1910. 1st edition. [x], 392pp. T.e.g. Front inner hinge tender. **$25.00\£14.00** *(BB 12/88)*

Ferrari, Giovanni Battista. Flora Quero Cultura di Fiori. Rome. Pier Ant. Facciotti, 1638. Small 4to, bound in full vellum (somewhat soiled). Red leather spine label, lacks 2 front end pages, engraved frontis, title with engraved border. [xii], 520pp, [xxvii] index. **$985.00\£547.00** *(MD 7/89)*

Fessenden, Thomas. Complete Farmer & Rural Economist. Boston, 1838. 345pp. Binding quite worn, contents water-stained. **$25.00\£14.00** *(AB 2/89)*

Fewster, Ernest. My Garden Dreams. Ottawa, 1926. 175pp. Near F. **$30.00\£17.00** *(Hrt 12/89)*

Finck, Henry. Gardening with Brains. Np, 1921. [xx], 270pp. F.f.e.p. loose, slightly tender else VG in 1/4 cloth on brds. **$16.00\£9.00** *(Hrt 12/89)*

Findlay, Hugh. Garden Making and Keeping. Garden City, 1926. 1st ed. Lg 8vo. [xviii], 252pp. VG. **$10.00\£6.00** *(RW 11/86)*

Findlay, Hugh. Garden Making and Keeping. Garden City, 1926. [xx], 252pp. Orig green clth, decor paper label. Some foxing, minor cover wear. About VG. **$25.00\£14.00** *(AT 3/89)*

Fink, B. The Lichen Flora of the U.S. A Reissue. MI, 1971. 426pp. VG in dj. **$20.00\£11.00** *(GW 8/89)*

Fink, Bruce. Lichens of Minnesota. Wash., GPO, USNH, 1910. 269pp. Wraps worn, else VG. **$30.00\£17.00** *(AB 4/89)*

Fiori, Adriano. Nuova Flora Analitica d'Italia. 1969-70. 3 vols. Hlf vellum type bndng. VG. **$81.00\£45.00** *(DL 89/90)*

Fischer, Helen. Flower Family Album. Minn., 1941. 130pp. VG. **$22.00\£12.00** *(AB 2/89)*
Another copy. U of Minn Press, 1941. 4to. 131pp. F in dj. **$12.50\£7.00** *(WH 1/90)*

Fisher, F.H. (ed). The Scree Garden. London, 1933. G. **$18.00\£10.00** *(ABx 3/89)*

Fisk, J. Success with Clematis. London, 1963. Reprint, 1962. 82pp. G in dj. **$22.50\£12.00** *(ABx 3/89)*

(Fithian) ed by H.D. Parish. Journals & Letters by Philip Vickers Fithian 1773-1774. Williamsburg, 1945. 3rd prntng. [xiv], 323pp. F in dj. **$40.00\£22.00** *(Hrt 12/89)*

(Fitton, Misses Elizabeth & Sarah May). Conversations on Botany. London, 1831. 7th ed. [xx], 278pp. Contemporary halfblue morocco and marbled brds. VG. **$234.00\£130.00** *(ABx 3/89)*

Fitzpatrick, Harry Morton. The Lower Fungi. Phycomycetes. NY, 1930. 1st ed. [xii], 322pp. VG. **$20.00\£11.00** *(JI 12/88)*

Flagg, Wilson. A Year Among the Trees. Boston, 1889. 308pp. Illus. Decorative green cloth. Heavily worn at foot of spine. **$14.00\£8.00** *(BB 12/88)*

Flemwell, G. Alpine Flowers and Gardens. London, 1910. 167pp. G. **$54.00\£30.00** *(ABx 3/89)*

Fletcher, David W. Himalayan Tea Garden. NY, 1955. 272pp. VG in wrn, reprd dj. **$15.00\£8.00** *(CB 1/90)*

Fletcher, H.L.V. Popular Flowering Shrubs. NY, 1972. 192pp. VG in worn dj. **$9.00\£5.00** *(CB 12/89)*

Fletcher, H.L.V. The Feature Garden. Branford, Mass, 1961. 1st Amer ed. [viii], 142pp. VG. **$10.00\£6.00** *(CB 2/90)*

Fletcher, H.L.V. The Fragrant Garden. London, 1965. 1st ed. 166pp. G in dj. **$11.00\£6.00** *(ABx 3/89)*

Fletcher, Harold & William Brown. The Royal Botanic Garden Edinburgh. London, 1970. 309pp. As New in dj. **$21.50\£12.00** *(ABx 3/89)*

Fletcher, Harold R. & William H. Brown. The Royal Botanic Garden, Edinburgh 1670-1970. Edinburgh, HMSO, 1970. 309pp. F in dj. **$30.00\£17.00** *(BF 2/90)*

Fletcher, S.W. How to Make a Fruit Garden. NY, 1914. [xix], 283pp. Repaired spine else VG. **$20.00\£11.00** *(RW 6/89)*

Fletcher, S.W. Strawberry Growing. NY, 1922. 325pp. VG. **$15.00\£8.00** *(AB 2/89)*

Flint, C. Grasses and Forage Plants. Crosby, 1860. 398pp. **$25.00\£14.00** *(GW 12/88)*

Flock, Elizabeth B. Wild Flowers of the Prairie Provinces. Regina, 1942. 112pp. Green leatherette (faded) wrappers. Signed by author and illustrator. G. **$9.50\£5.00** *(BB 12/88)*

Fogg, H.G. Witham. The Book of Bulbous Plants. London, 1957. 224pp. Black cloth. VG in dj. **$9.50\£5.00** *(BB 12/88)*

Fogg, J. Jr. The Flora of the Elizabeth Islands. Mass, 1930. 162pp. Gray. Wrs., signed. VG. **$10.00\£6.00** *(GW 8/89)*

Foley, Dan. Complete Book of Garden Ornaments. NY, 1972. Nearly New in dj. **$20.00\£11.00** *(AB 4/89)*

Another copy. NY, 1972. 247pp. VG in repaired dj. **$19.00\£11.00** *(CB 12/89)*

Another copy. NY, 1972. 247pp. Green cloth, small piece out of rear end page else VG in worn dj. **$15.00\£8.00** *(MD 2/90)*

Foley, Dan. Flowering World of "Chinese" Wilson. NY, 1969. 334pp. As New in sun faded dj. **$12.00\£7.00** *(AB 4/89)*

Another copy. London, 1969. 1st ed. 334pp. VG. **$20.00\£11.00** *(RW 6/89)*

Another copy. NY, 1969. 1st prntng. 334pp. F in dj. **$15.00\£8.00** *(WH 1/90)*

Foley, Dan. The Complete Book of Garden Ornaments, Complements, and Accessories. NY, 1972. 247pp. F in slightly wrn dj. **$35.00\£19.00** *(BF 2/90)*

Foley, Dan. The Flowering World of "Chinese" Wilson. London, 1969. [xv], 334pp. Almost As New in dj. **$40.00\£22.00** *(Hrt 12/89)*

Fontana, F. Observations on the Rust of Grain. 1767 (1932; Phytopath. Classics #2). Wrps. G. **$10.00\£6.00** *(PBx 11/89)*

Foote, H.R. Mrs. Foote's Rose Book. Boston, 1948. Nice, crisp copy. **$14.00\£8.00** *(PBx 11/89)*

Forrest, G. George Forrest. London, 1935. 89pp. Scarce. G. **$57.50\£32.00** *(ABx 3/89)*

Forrest, G. Some Plants, Shrubs and Trees Found in 1925. London, 1926. 126pp. Priv prnted. G. **$36.00\£20.00** *(WW 1/90)*

Forrest, George. (ed). by Dr. J. Macqueen Cowan. Journeys and Plant Introductions of George Forrest V.M.H. London, 1952. [xi], 252pp. Nr F in dj. Scarce. **$80.00\£44.00** *(Hrt 12/89)*

Forsyth, W. A Treatise on the Culture and Management of Fruit Trees. . . to Which Are Added an Introduction and Notes Adapting. . . The Treatise to the Climates and Seasons of the USA by Wm. Cobbett. Albany, NY, 1803. More recent bndng; perfect. Contnts some brwning & age sptting. Plts reprd by previous ownr in 1834. Attractive. **$155.00\£86.00** *(PBx 11/89)*

Another copy. London, 1803. 2nd ed. [xxvii], 523pp. Clf; neatly rebckd. G. **$99.00\£55.00** *(WW 1/90)*

Forsyth, W. Treatise on Fruit Trees. 1824. 7th and last ed. Bnd in 1/2 green morocco. VG. **$90.00\£50.00** *(DL 89/90)*

Forsyth, William. A Treatise on the Culture and Management of Fruit Trees. London, 1806. 4th ed. (1802). 523pp. Rebound in quarter brown calf with marbled boards. Lovely copy. **$171.00\£95.00** *(ABx 3/89)*

Fortune, Robert. A Residence among the Cinese. London 1857. [xvi], 440pp. Early, very wrn hlf lethr. F.e.p.s, 1st & last page brwnd/smudged. Else G clean cpy. Scarce. **$290.00\£161.00** *(Hrt 12/89)*

Foster, Gertrude. Herbs for Every Garden. NY, 1966. 255pp. VG in sunned dj. **$10.00\£6.00** *(AB 4/89)*

Foster, H. Lincoln. Rock Gardening. Boston, 1968. 1st ed. [xxviii], 466pp. VG in dj. **$20.00\£11.00** *(RW 11/86)*

Another copy. Boston, 1968. 1st ed. [xxviii], 466pp. VG in worn dj. **$20.00\£11.00** *(CB 12/89)*

Foster-Melliar, Rev. A. The Book of the Rose. London, 1894. 1st ed. 336pp. Slightly worn. **$14.50\£8.00** *(ABx 3/89)*

Fowells, H. Silvics of Forest Trees of the U.S. USDA, 1965. 762pp. VG. **$30.00\£17.00** *(GW 12/88)*

Fox, Frances M. Flowers and Thier Travels. NY, 1936. 229pp. VG. **$12.50\£7.00** *(RW 6/89)*

Fox, Helen M. Adventure in My Garden. NY, 1965. 167pp. VG. **$20.00\£11.00** *(RW 6/89)*

Fox, Helen Morgenthau. Adventure in My Garden. NY, 1965. 167pp. VG in slightly chppd dj. **$25.00\£14.00** *(BF 2/90)*

Fox, Helen Morgenthau. Andre le Notre, Garden Architect to Kings. NY, 1962. 176pp. F in slightly chppd dj. **$30.00\£17.00** *(BF 2/90)*

Fox, Helen Morgenthau. Garden Cinderellas. NY, 1928. 269pp. Slightly loose else G. **$25.00\£14.00** *(ABx 3/89)*

Fox, Helen Morgenthau. Gardening with Herbs. NY, 1970. Facsimile of 1933 ed. 334pp. VG. **$7.50\£4.00** *(RW 11/86)*

Another copy. NY, 1934. 334pp dec cloth in torn dj. G. **$43.00\£24.00** *(ABx 3/89)*

Another copy. NY, 1933. 1st ed. [xiv], 334pp. Label missing else VG. **$25.00\£14.00** *(CB 12/89)*

Fox, Helen Morgenthau. Patio Gardens. NY, 1929. Internally near F; yellow clth somewhat soiled. **$65.00\£36.00** *(BF 2/90)*

Foyle, W.J. Culpeper's English Physician and Complete Herbal. Foyle, 1932. 139pp. VG. **$12.00\£7.00** *(GW 12/88)*

Fraser, Hugh. Handy Book of Ornamental Conifers. Edinburgh, 1875. 1st ed. 292pp. G. **$29.00\£16.00** *(ABx 3/89)*

Fraser, S. American Fruits. NY, 1924. 888pp. Cloth worn, contents reinforced in binding else VG. **$20.00\£11.00** *(AB 2/89)*

Freed, Hugo. Orchids and Serendipity. Englewood Cliffs, 1970. 1st ed. [viii], 184pp. VG in dj. **$15.00\£8.00** *(JI 12/88)*

French, Henry. Farm Drainage. NY, 1880. 384pp. G. **$15.00\£8.00** *(CB 1/90)*

Friesner, Ray. Key to Species & Varieties of Ferns & Fern Allies of N.E. North America. Butler U., 1940. 21pp. Wraps. VG. **$8.00\£4.00** *(AB 4/89)*

Frohman, Louis & Jean Elliot. A Pictorial Guide to American Gardens. NY, 1960. 364pp. G in torn dj. **$27.00\£15.00** *(ABx 3/89)*

Frye, T. Ferns of the Northwest. Metro, 1934. 177pp. VG. **$15.00\£8.00** *(GW 8/89)*

Fukuda, Kazuhiko. Japanese Stone Gardens/ How to Make and Enjoy Them. Rutland, 1970. 312pp. F in dj. **$40.00\£22.00** *(BF 2/90)*

Fulford, M. The Genus Bazzannia in Central and South America. Chronica Botanica, 1946. 175pp. VG. **$30.00\£17.00** *(GW 12/88)*

Fuller, A.S. Illustrated Strawberry Culturist. NY, 1895. (1887). 59pp. Stiff boards. Almost new. **$15.00\£8.00** *(AB 4/89)*

Fuller, Andrew S. The Small Fruit Culturist. NY 1867. 276pp. Wrn gilt clth (as usual) else VG. **$55.00\£31.00** *(Hrt 12/89)*

Fuller, Andrew. The Grape Culturist. NY, 1886. New & enlrgd. vi, 7-286pp. (2) advts. Orig glt brwn clth. VG. **$40.00\£22.00** *(Hrt 12/89)*

Fuller, Andrew. The Nut Culturist. NY, 1919. (1896). [viii], 289pp. VG. **$19.50\£11.00** *(BB 12/88)*

Fullerton, Edith L. How to Make a Vegetable Garden. NY, 1905. [xix], 347pp. VG. **$20.00\£11.00** *(RW 6/89)*

Fultz, Francis. Lily, Iris and Orchid of Southern California. Gardena, 1928. Trade ed. 135pp. Front endpaper out, spine print dull, rare. **$22.50\£13.00** *(BLB 7/89)*

G

Gadeau de Kerville, Henri. Notes sur le Fourgeres. Rouen 1915, 1917, 1922. Parts #1-#11. 3 extracted pieces. 149pp. Original wrappers. VG set. **$17.50\£10.00** *(WH 1/90)*

Galbally, John. The Complete Guide to Border Carnations. London, 1966. 1st ed. 128pp. Fine in dj. **$14.00\£8.00** *(BB 12/88)*

Gallardo, D.I. Perez. Manual de Agricultura y Ganaderia Mejicanas. Paris, 1862. 2nd ed. [vi], (2), 9-288pp. 1/2 clth/mrbld brds, showing wear else quite G. **$30.00\£17.00** *(Hrt 12/89)*

Galloe, O. Natural History of the Danish Lichens. Copenhagen, 1927-72. 1397 plates, some colored, of 7004 figs. 10 vols. Orig wrappers. G. **$216.00\£120.00** *(WW 1/90)*

Galloway, B.T. Commercial Violet Culture. NY, 1899. 12mo. 224pp. G only. **$8.00\£4.00** *(CB 12/89)*

Gamble, J.A. Roses Unlimited. Wash., 1950. Priv. publ. by author. 82pp. F. **$20.00\£11.00** *(PBx 11/89)*

(Garden Club of Virginia) A Garden Potpourri from the Garden Club of Virginia Journal. Richmond, 1963. 331pp. F in dj. **$15.00\£8.00** *(BF 2/90)*

Garden Magazine. Vol. I: Feb-July 1905. Vol II: Aug 1904-Jan 1906. Blue cloth. 2 vols. **$65.00\£36.00** *(RW 6/89)*

Gardiner, W. The Flora of Forfarshire. London & Edinburgh, 1848. [xxiv], 308pp. 12mo. New cloth. Stmp on title and reverse of frontis. **$99.00\£55.00** *(WW 1/90)*

Gardner, C.A. West Australian Wild Flowers. Perth, 1947. 6th ed. 4to. G. **$27.00\£15.00** *(WW 1/90)*

Gardner, C.A. Wildflowers of Western Australia. Perth, 1959. [xii], 164pp with map. Green cloth over boards with internal spiral binding. Sm 4to. VG in dj. **$33.00\£18.00** *(BB 12/88)*

Gardner, N.L. The Genus Fucus on the Pacific Coast of North America. Berkeley, 1922. 180pp. Ry 8vo, wrpprs. G. **$36.00\£20.00** *(WW 1/90)*

Gardner, Rose. The Garden Anthology. London, 1910. Reprint, 1906. 311pp. Pink calf very worn. Reading copy only. **$9.00\£5.00** *(ABx 3/89)*

Gardner, V.R. et al. Orcharding. NY, 1927. 311pp. Cloth and contents VG. **$22.50\£13.00** *(AB 2/89)*

Garner, R.J. The Grafter's Handbook. London, 1950. 3rd imp. 224pp. G in dj. **$14.50\£8.00** *(ABx 3/89)*

Garrard, Jeanne & Hannau, Hans. Flowers of the Bahamas. Miami, c1975. Oblong small 12mo. Pictorial boards, 64pp. F. **$5.75\£3.00** *(BLB 7/89)*

Garrett, G. Spring Flora of the Wasatch Region. Stevens, 1936. 241pp. VG. **$15.00\£8.00** *(GW 8/89)*

Garris & Hoffmann. Southern Horticultural Enterprises. Phila., 1938. 551pp. VG. **$12.00\£7.00** *(AB 2/89)*

Gates, F. Wild Flowers in Kansas. KS, 1934. 295pp. Wraps. Taped spine else VG. **$10.00\£6.00** *(GW 12/88)*

Gathorne-Hardy, R. The Tranquil Gardener. London, 1958. Orig clth. G. **$36.00\£20.00** *(WW 1/90)*

(Gatty, Mrs. Alfred) edited by Eden, H.K.F. & E. Lloyd. The Book of Sun Dials. London, 1900. 4th ed. [xvii], 529pp. VG in gilt dec red clth. **$165.00\£92.00** *(Hrt 12/89)*

Gatty, Mrs. Alfred. The Book of Sundials. London, 1900. 4th ed., red clth, t.e.g. 529pp. Shlf wrn else VG. **$120.00\£67.00** *(MD 2/90)*

Gault, S. Millar & Patrick Synge. The Dictionary of Roses in Color. NY, 1971. [xlv], 191pp. 4to. Fine in dj. **$33.00\£18.00** *(BB 12/88)*
Another copy. London, 1971. [xlv], 191pp. Pbck. G. **$18.00\£10.00** *(ABx 3/89)*

Gaumann, E. & C.W. Dodge. Comparative Morphology of Fungi. NY, 1928. [xiv], 701pp. G. **$36.00\£20.00** *(WW 1/90)*

Gauthier, Joseph-Stany. Les Maisons Paysannes des Vielles Provinces de France. Paris, nd, pre-1945. 235pp. VG in dj. **$55.00\£99.00** *(Hrt 12/89)*

Gauthier-Lievre, L. Zygnemacees Africaines. Beth. Nov. Hedw. 20, 1965. [viii], 210pp. Ry 8vo, wrpprs. G. **$36.00\£20.00** *(WW 1/90)*

Gautier, L.M. Les Champignons. Paris, 1884. [xvi], 508pp. Orig dec clth, trifle wrn. Plts slightly fxed. **$180.00\£100.00** *(WW 1/90)*

Geld, Ellen Bromfield. The Heritage. NY, 1962. 204pp. VG in worn dj. Scarce. **$12.00\£7.00** *(AB 2/89)*

Genders, Roy. Anemones. London, nd (1963?). 123pp. Garden Book Club. G in dj. **$12.50\£7.00** *(ABx 3/89)*

Genders, Roy. Bedding Plants. London, 1956. 192pp. VG. **$7.50\£4.00** *(RW 6/89)*

Genders, Roy. Colour All the Year Round. NY, 1963. 1st Amer ed. 287pp. VG in reprd dj. **$12.00\£7.00**

Genders, Roy. Garden Pinks. London, 1962. 1st ed. 160pp. VG in slightly chipped dj. **$18.50\£10.00** *(BB 12/88)*

Genders, Roy. Miniature Roses. London, 1960. 104pp. VG in dj. **$15.00\£8.00** *(AB 4/89)*

Genders, Roy. The Cottage Garden. London, 1970. 1st ed. 2nd imp. 368pp. G in dj. **$36.00\£20.00** *(ABx 3/89)*

Genders, Roy. The Rose. London, 1965. 1st ed. 623pp. G in repaired dj. **$36.00\£20.00** *(ABx 3/89)*

Gerarde, J. Leaves from Gerard's Herball. London, 1931. [xii], 305pp. G. **$32.00\£18.00** *(WW 1/90)*

Gibbons, Euell. Stalking the Faraway Places. NY, 1973. 279pp. VG in edgeworn dj. **$10.00\£6.00** *(AB 4/89)*

Gibbs, Rebecca. A Garland of Essays. Berkshire Hills, MA, 1932. Ltd to 100 copies. 107pp. As New. **$18.00\£10.00** *(AB 2/89)*

Gibson, J.L. Carnations for Amateurs. London, 1953. 3rd rev. ed. 136pp. VG-F in G dj. **$7.75\£4.00** *(BLB 7/89)*

Gibson, W. Eye Spy; Afield with Nature among Flowers and Animate Things. Harper, 1900. 264pp. **$25.00\£14.00** *(GW 12/88)*

Gibson-Carter, H. Glossary of the British Flora. Cambridge, 1950. 79pp. VG in dj. **$9.00\£5.00** *(GW 12/88)*

Gilkey, Helen. Weeds of the Pacific Northwest. Oregon State, 1957. [xxix], 441pp. G. **$19.00\£11.00** *(CB 1/90)*

Gillespie, Janet. The Joy of a Small Garden. NY, 1963. [viii], 230pp. VG in dj. **$15.00\£8.00** *(Hrt 12/89)*

Gillespie, Norvell. A Pacific Coast Gardening Guide. Garden City, NY, 1949. [x], 306pp. Near fine in chipped dj. **$9.00\£5.00** *(BB 12/88)*

Gilmour, John & Max Walters. Wild Flowers. London, 1969. 4th ed. [xiv], 242pp. VG. **$12.00\£7.00** *(Hrt 12/89)*

Gilmour, John. British Botanists. London, 1946. 48pp. VG in dj. **$7.50\£4.00** *(RW 11/86)*

Gilmour, John. Wild Flowers of the Chalk. London, 1947. 1st ed. 32pp. Small 8vo. King Penguin #37. G. **$11.00\£6.00** *(ABx 3/89)*

Gilpin, William. Observations on the Western Parts of England. London, 1798. 1st ed. [xvi], 359pp. (1) adv. hlf lthr, gilt title, some rbbng, bookplt, pencil notes f.f.e.p. else very nice cpy. **$245.00\£136.00** *(Hrt 12/89)*

Givens, Harold D. The Secrets of Artistic Gardens. NY, 1969. 131pp. VG in dj. **$32.00\£18.00** *(Hrt 12/89)*

Gleason, Henry A. & Arthur Cronquist. Manual of Vascular Plants of Northeastern United States and Adjacent Canada. Princeton, 1963. [li], 810pp. Green cloth. Neat ex-library. No external markings. VG. **$24.00\£13.00** *(BB 12/88)*

Goddard, J.H. The New Chrysanthemum Treasury. London, 1953. 2nd ed. 258pp. G. **$11.00\£6.00** *(ABx 3/89)*

Godwin, H. The History of the British Flora. Cambridge U Press, 1956. 26 plts, 119 other illus, figs & maps, bibl., fldg table. Near F. **$36.00\£20.00** *(PBx 11/89)*

Goebel, Friedemann & Gustav Kunze. Pharmaceutische Waarekunde. Eisenach, 1827-34. 2 vols in one. [viii], [iv], 240pp; [viii], 300pp. Near F in library binding. Ex-lib. Rare. **$720.00\£400.00** *(Hrt 12/89)*

Goldberg, A. The Genus Melochia L. (Sterculiaceae). Contrib. U.S. Nat. Herb. 34 (5). 1967. Pbnd. 174pp. G. **$11.00\£6.00** *(PBx 11/89)*

Goldring, William. The Book of the Lily. London, 1905. [viii], 98pp. Green cloth. Light foxing, G. **$18.00\£10.00** *(BB 12/88)*

Goldsmith, Margaret O. Designs for Outdoor Living. NY, 1941. [xxv], 358pp. Pres cpy from authr. BF. **$45.00\£25.00** *(Hrt 12/89)*

Goldsmith, Margaret O. Designs for Outdoor Living. NY, 1941. 1st ed. [xxv], 358pp. VG. **$15.00\£8.00** *(RW 11/86)*

Gomez, Natalie. You Garden in the City. NY, 1941. [xiv], 247pp. VG in dj. **$22.00\£12.00** *(Hrt 12/89)*

Gomont, M. Monographie des Oscilariees. Paris, 1892-93. (1962). 302pp. G. **$27.00\£15.00** *(WW 1/90)*

Good, Peter P. The Family Flora and Materia Medica Botanica. Vol. I: Cambridge, MA, 1854. Vol. II: Elizabethtown, NJ, 1845. New ed., revd. and enlrgd. Publisher's dec cloth, some plates discolored upper left hand corner. Small piece missing f.f.e.p., top of spine chipped else VG. Uncommon. **$440.00\£244.00** *(Hrt 12/89)*

Good, R. d'O. The Geography of Flowering Plants. London, 1964. 3rd ed. G. **$27.00\£15.00** *(WW 1/90)*

Good, Ronald. The Geography of the Flowering Plants. London, 1953. 2nd ed. 452pp. F in dj. **$27.50\£15.00** *(WH 1/90)*

Good, William. Garden Work. NY, nd (c1913). [xvi], 359pp. Dec cvr. VG. **$20.00\£11.00** *(RW 11/86)*

Goodrich, Charles. The First Book of Farming. NY, 1923. [xx], 259pp. VG. **$7.50\£4.00** *(RW 6/89)*

Goodspeed, T. Harper. Plant Hunters in the Andes. NY, 1941. [xvi], 429pp. Lge 8vo. VG in dj. **$45.00\£25.00** *(RW 11/86)*

Another copy. NY, 1941. 429pp. G in dj. **$81.00\£45.00** *(ABx 3/89)*

Another copy. CA, 1961. 378pp. Photos. Spine faded else VG. **$26.00\£14.00** *(GW 8/89)*

Another copy. Berkeley, 1961. 2nd, rev. ed. [x], 380pp. VG in dj. **$35.00\£19.00** *(JI 12/88)*

Gopalaswamiengar, K.S. Complete Gardening in India. Bangalore, 1935. Rev. ed. 639pp. VG in dj. **$10.00\£6.00** *(WH 1/90)*

Gordon, J. Pageant of the Rose. 1953. 1st ed. F. **$20.00\£11.00** *(PBx 11/89)*

Gordon, Jean. Pageant of the Rose. Woodstock, 1961. 2nd rev ed. 232pp. Some lght spttng to covrs else VG in tattered dj. **$17.00\£9.00** *(CB 1/90)*

Gordon, Lesley. Poorman's Nosegay—Flowers from a Cottage Garden. London, 1973. 222pp. Slightly rbbd bndng, top edge bmpd. **$35.00\£19.00**

Gorer, Richard. Living Tradition in the Flower Garden. Newton Abbot, 1974. 242pp. F in dj. **$25.00\£14.00** *(BF 2/90)*

Gorer, Richard. The Development of Garden Flowers. London, 1970. 254pp. F in dj. **$20.00\£11.00** *(BF 2/90)*

Gorer, Richard. The Development of Garden Flowers. London, 1970. 1st ed. 254pp. Scarce. Fine in dj. **$28.00\£16.00** *(BB 12/88)*

Gorer, Richard. The Flower Garden in England. London, 1975. 178pp. F in dj. **$25.00\£14.00** *(BF 2/90)*

Gothein, Mary Luise. A History of Garden Art. London, 1928. 1st English ed. 2 vols. [xxiv], 459pp; [xv], 486pp. Scarce edition. VG. **$378.00\£210.00** *(ABx 3/89)*

Gould, F. The Grasses of Texas. TX A&M, 1975. 653pp. 330 fig. VG in dj. **$20.00\£11.00** *(GW 8/89)*

Grabham, M.C. The Garden Interests of Madeira. London, 1926. Cr 8vo. Slightly fxed. G. **$27.00\£15.00** *(WW 1/90)*

Graf, Alfred Byrd. Exotica. Series 3. E. Rutherford, 1973. 6th ed. 1834pp. 4to. VG. **$112.00\£62.00** *(BB 12/88)*

Graham, Edward. Legumes for Erosion Control and Wildlife. GPO, 1941. 153pp. Stamps on endpapers else VG. **$15.00\£8.00** *(CB 12/89)*

Graham, V.E. Tropical Wild Flowers. London, 1963. 200pp. Orange cloth. VG in dj. **$14.50\£8.00** *(BB 12/88)*

Gram, E. & A. Weber. Plant Diseases in Orchard, Nursery and Garden Crops. London, 1952. 618pp. Two lves slightly affected by adhesion. G. **$36.00\£20.00** *(WW 1/90)*

Gram, E. & A. Weber. Växetsjukdomar. Stockholm, 1942. Ry 8vo. G. **$21.00\£12.00** *(WW 1/90)*

Grant, V. Flower Pollination in the Phlox Family. Columbia, 1965. 180pp. VG in dj. **$15.00\£8.00** *(GW 8/89)*

Grantwhite, J.E. Garden Art and Architecture. NY, 1969. [xxxi], 110pp. Almost As New in dj. **$45.00\£25.00** *(Hrt 12/89)*

Grasby, Nancy. Imaginative Small Gardens. NY, 1963. 256pp. VG in dj. **$10.00\£6.00** *(RW 6/89)*

Graves, Arthur Harmount. Illustrated Guide to Trees and Shrubs. Wallingford, CT, 1952. [x], 240pp. Privately published. Torn/mended dj. **$12.50\£7.00** *(BB 12/88)*

Graves, C. Catalog of the Flowering Plants & Ferns of Connecticut. CT, 1910. 569pp. Wrs. VG. **$25.00\£14.00** *(GW 8/89)*

Graves, Ch. B et al. Catalogue of the Flowering Plants & Ferns of Connecticut. Hartford, 1910. 569pp. Pbnd. G. **$16.00\£9.00** *(PBx 11/89)*

Gray, Asa. Gray's School and Field Book of Botany. NY, 1868. [xii], 236pp; 386pp. Consisting of "Lessons in Botany" and "Field, Forest and Garden Botany" bound in 1 volume. Marbled edges. Shelf wear on lower extremities. **$23.00\£13.00** *(BB 12/88)*

Gray, Asa. Manual of the Botany of the Northern United States. 6th ed. 1889. 760pp. Sft lthr bndng. Some rbbng, overall excellent. **$25.00\£14.00** *(PBx 11/89)*

Gray, Asa. New Manual of Botany—A Handbook of the Flowering Plants & Ferns. . . 7th ed. 1908. Fll flex. Lthr, thin paper. 926pp. Contents F, bndng poor & flaky. **$15.00\£8.00** *(PBx 11/89)*

(Gray, Asa) Scientific Papers of Asa Gray. Sel. by Ch. S. Sargent. 2 vols in 1. 1889 (1969). 900pp. F. **$24.00\£13.00** *(PBx 11/89)*

Gray, H.L. The English Field System. Harvard, London, 1959. 568pp. VG in trn dj. **$20.00\£11.00** *(CB 1/90)*

Gray's Manual of Botany. NY, 1950. 8th ed. [lxiv], 1632pp. VG. **$40.00\£22.00** *(WH 1/90)*

Graydon, N. South Carolina Gardens. Beaufort, 1973. 78pp. VG in dj. **$18.00\£10.00** *(GW 12/88)*

Grayson, Dave. Countryman's Year. GC, 1936. 270pp. As new in dj. **$5.00\£3.00** *(AB 2/89)*

Grayson, Dave. Under my Elm. GC, 1942. 278pp. As new in dj. **$5.00\£3.00** *(AB 2/89)*

Green, C. The Flora of the Liverpool District. Buncie, 1933. 163pp. Name sticker on flyleaf, else VG. **$35.00\£19.00** *(GW 8/89)*
Another copy. Arbroath, 1933. [xi], 163pp. Trifle used. **$27.00\£15.00** *(WW 1/90)*

Green, S.B. Popular Fruit Growing. 5th ed. St. Paul, 1918. 328pp. VG. **$15.00\£8.00** *(AB 2/89)*

Greene, H. and Blomquist, W. Flowers of the South: Native and Exotic. NC, 1973. 208pp. VG in dj. **$11.00\£6.00** *(GW 12/88)*

Gremli, A. The Flora of Switzerland. Zurich, 1888. Translated into English by Leonard Paitson from the 5th German ed. 454pp. G. **$27.00\£15.00** *(ABx 3/89)*

Grieve, M. Culinary Herbs and Condiments. NY, 1934. 1st U.S. 209pp. Owner's stamps and plate, else VG $30.00\£17.00 *(Hrt 12/89)*

Griffith, Mary S. Gardening on Nothing a Year. Boston, 1937. 231pp. Ownr's sig, slightly soiled clth else VG. **$15.00\£8.00** *(BF 2/90)*

Grigson, G. A Dictionary of English Plant Names. London, 1974. 254pp. G. **$27.00\£15.00** *(WW 1/90)*

Grigson, Geoffery. Wild Flowers in Britain. London, 1948. 48pp. VG in dj. **$7.50\£4.00** *(RW 6/89)*

Grigson, Geoffrey. Flowers of the Meadow. London, 1950. 1st ed. 36pp. G in dj King Penguin #53. **$16.00\£9.00** *(ABx 3/89)*

Grigson, Geoffrey. The Englishman's Flora. London, 1955. 1st ed. 478pp. 4to. G. **$100.00\£55.00** *(ABx 3/89)*

Grimwade, R. An Anthology of the Eucalypts. Sydney, 1920. Orig brds; clth bck. G. **$72.00\£40.00** *(WW 1/90)*

Grisebach. A. La Vegetation Du Globe. Paris, 1877-76. 2 vols. Ry 8vo. Mdrn clth, G. **$108.00\£60.00** *(WW 1/90)*

Grohmann, J.G. & Baumgarten, F.G. Ideenmagazin für Liebhaber von Garten, Englischen Anlagen. Leipzig. Baumgaertner. 5 vols bnd in 4. Vol. I: 1799, 3rd ed. Vol II: 1800, 3rd ed. Vol III: 1797, 2nd ed. Vol IV: the 2 French/German title pp mssng. Vol V: 1806. Lg 4to, 546plts, (as called for), 59 hand colored. There are 60 cahiers, 12 per vol. Cahiers contain 7 to 10 plts each and are preceded with explanatory text in French/German. An index for each vol, except III, in French/German. B/w plts frun from very clean to fxed, heaviest on mrgns. Water stn 2" into 8 plts. Hand colored plts very crisp & clean. Bnd in early hlf lhr on mrbld brds. Dec spine, mssng 1 title label. 1 vol # chppd. Frnt brd of one vol loose. Complete set exceedingly rare. Exceptionally G cpy. **$14,500.00\£8056.00** *(Hrt 12/89)*

Gronovius, J. Flora Virginica. Arnold, 1946. 202pp. Wraps, VG. **$15.00\£8.00** *(GW 12/88)*

Grontved, J. et al (eds). The Botany of Iceland – The Pteridophyta and Spermaphyta of Iceland. 1942. Paper cvrs as issued. **$45.00\£25.00** *(DL 89/90)*

Grootendorst, H.J. Rhododendrons En Azaleas. Boskoop, 1954. 216pp. Tall 8vo. G. **$34.00/£19.00** *(PBx 11/89)*

Grout, A.J. Mosses with a Hand-Lens. NY, 1924. 3rd ed. [xv], 339pp. Sl tender, stned brds, ink notes inside bck cvr. **$18.00\£10.00** *(Hrt 12/89)*

Grout, A.J. Mosses With Hand-Lens and Microscope. NY, 1903. 1st ed. Signed. 416pp. Glt stmpd spine, shows some shlf wr. **$85.00\£47.00** *(Hrt 12/89)*

Grove, A. Lilies. London & Edinburgh, nd (1922). 116pp. Apart from slight chipping at head and tail, a VG copy of one of the scarcer titles in this collect series. **$21.50\£12.00** *(ABx 3/89)*

Grove, W.B. The British Rust Fungi. Cambridge, 1913. Trifle used. **$27.00\£15.00** *(WW 1/90)*

Groves, J. Walton. Edible and Poisonous Mushrooms of Canada. Ottawa, 1962. [x], 298pp. VG. **$30.00\£17.00** *(Hrt 12/89)*

Grubb, Norman H. Cherries. London, 1949. 1st ed. 189pp. VG. **$63.00\£35.00** *(ABx 3/89)*

Guiness, Desmond & William Ryan. Irish Houses & Castles. NY, 1971. 371 plts. Sm folio. Almost As New in dj. **$65.00\£36.00** *(Hrt 12/89)*

Gunther, R.T. Early British Botanists and Their Gardens. 1922 (1971). 417pp. F. **$32.00\£18.00** *(PBx 11/89)*

Gunther, R.T. Early English Botanists and Their Gardens. NY, 1971. Kraus reprint. 417pp. G. **$54.00\£30.00** *(ABx 3/89)*

Gunther, R.T. Oxford Gardens. Oxford, 1912. 280pp. VG. **$72.00\£40.00** *(ABx 3/89)*

Guppy, H.B. Observations of a Naturalist in the Pacific between 1896 and 1899. London. 1st ed. 2 vols. Vol. I: 1903, [xxx], 627pp. Vol. II: 1906, [xxviii], 626pp. Sm libry stmp on title pp & few others, almost as never used. Nice set. Scarce. **$300.00\£167.00** *(Hrt 12/89)*

Gussone, J. Enumeratio Plantarum Vascularium in Insula Inarime Sponte Provenientium. Naples, 1855. [iv], [xix], 428pp,[7]. Some very lt fxing. Stmp on title pge. New qurter clf. G. **$225.00\£125.00** *(WW 1/90)*

Gussow, H.T. & W.S. Odell. Mushrooms and Toadstools. Ottawa, 1927. 274pp. Glt stmpng. Spine rbbd, soiled. G. **$30.00\£17.00** *(CB 1/90)*
Another copy. 1927. Nice cond. **$35.00\£19.00** *(PBx 11/89)*
Another copy. 1927. VG. **$36.00\£20.00** *(DL 89/90)*

Gussow, H.T. & W.S. Odell. Mushrooms and Toadstools. Ottawa, 1927. 1st ed. [vi], 276pp. Sm 4to. Usual fading on spine, else VG. **$45.00\£25.00** *(JI 12/88)*

Gwynne-Vaughan, H.C.I. & B. Barnes. The Structure and Development of the Fungi. Cambridge, 1930. [xvi], 384pp. Some shlfwr else VG. **$18.00\£10.00** *(Hrt 12/89)*

H

Haddington, 6th Earl of. Forest Trees. London, 1953. 95pp. Chipped dj. G. **$29.00\£16.00** *(ABx 3/89)*

Haddington, Earl of. Forest Trees. London, 1953. [xlii], 95pp. G. **$32.00\£18.00** *(WW 1/90)*

Hadfield, M. & J. Gardens of Delight. London, 1964. 192pp. G. **$36.00\£20.00** *(WW 1/90)*
Another copy. Boston, 1964. 1st Amer ed. 192pp. Rev of f/p slightly fxd else VG. **$38.00\£21.00** *(BF 2/90)*

Hadfield, Miles (ed). The Gardener's Album. London, 1954. Large 8vo. 175pp. VG in dj. **$25.00\£14.00** *(RW 11/86)*

Hadfield, Miles (ed). The Gardener's Companion. London, 1936. 1st ed. 624pp. G. **$25.00\£14.00** *(ABx 3/89)*

Hadfield, Miles & John. Gardens of Delight. NY, 1964. 1st US ed. 192pp. G. **$27.00\£15.00** *(ABx 3/89)*

Hadfield, Miles, E.A. Bunyard et al. The Gardener's Companion. London, 1936. 1st ed. [xvi], 623pp. VG in wrn dj. **$28.00\£16.00** *(CB 1/90)*

Hadfield, Miles. Gardens. NY, 1962. 128pp. VG in box. **$15.00\£8.00** *(RW 11/86)*

Hadfield, Miles. Pioneers in Gardening. London, 1955. 240pp. Internally F. Top of cover sunned, dj slightly chppd. **$30.00\£17.00** *(BF 2/90)*

Hadfield, Miles. Pioneers of Gardening. London, 1955. 1st ed. [xiii], 240pp. VG in worn dj. **$21.00\£12.00** *(CB 12/89)*

Hadfield, Miles. The Art of the Garden. London, 1965. 160pp. Wraps. VG. **$9.00\£5.00** *(RW 6/89)*

Hadfield, Miles. Topiary and Ornamental Hedges. London, 1971. 100pp. G in dj. **$36.00\£20.00** *(ABx 3/89)*

Hadlington, P & Staunton, I. Shrubs and Trees for Austrailian Gardens. Sydney, 1973. 1st, 1g 8vo. 106pp. F in dj. **$11.75\£7.00** *(BLB 7/89)*

Haes, E.C.M. Winter Colour in the Garden. London, 1965. Garden Book Club Edition, 104pp. G in dj. **$9.00\£5.00** *(ABx 3/89)*

Hagerup, V. & Peterson, O. Botanical Atlas. Munk, 1959. 2 vols. 848pp. 1/2 leather. Shelfwear on bottom of boards. In Swedish & English. VG. **$65.00\£36.00** *(GW 8/89)*

Haggard, H. Rider. A Gardener's Year. London, 1905. 404pp. VG. **$75.00\£42.00** *(RW 6/89)*

Haines, H.H. The Botany of Bihar and Orissa. London, 1921-25. [x], 199pp. 3 vols. rare orig. prntng. G. **$54.00\£30.00** *(WW 1/90)*

Haines, Jennie Day (ed). Ye Gardeyne Boke, A Collection of Quotations Instructive and Sentimental. SF, 1906. 72pp. Pttrnd clth; edge wear else VG. **$35.00\£19.00** *(BF 2/90)*

Hale, M. The Biology of Lichens. Arnold, 1974. 181pp. Wraps. VG. **$9.00\£5.00** *(GW 12/88)*

Hale, S.J. Flora's Interpreter: or, The American Book of Flowers and Sentiments. Boston, 1842. 262pp. 11th ed. Orig clth, trifle rbbed. Frnt jnt sm crck. G. **$49.00\£27.00** *(WW 1/90)*

Hall, H. The Genus Haplopappus:A Phylogentic Study in the Compositae. Carnegie, 1928. 391pp. Wraps. VG. **$20.00\£11.00** *(GW 12/88)*

Hall, H.M. & C.C. Hall. A Yosemite Flora. SF, 1912. [vii], 282pp. Pblshr's insert–"this cpy lacks plates 2 to 11, o/w complete". Scarce. VG. **$14.00\£8.00** *(Hrt 12/89)*

Hall, Harvey M. & T.H. Goodspeed. A Rubber Plant Survey of Western N. America. CA, 1919. 119pp. Wraps. Signed. VG. **$10.00\£6.00** *(GW 12/88)*

Hambidge, Gove. Hunger Signs in Crops. WA., Amer. Soc. Agron., 1950. 3rd ed. 350pp. Light wear. **$15.00\£8.00** *(AB 2/89)*

Hamer, A. Handel. Wild Flowers of the Cape; A Floral Year. Cape Town, n.d. (1926). 104pp. 4to. Green cloth. Some light foxing. RFEP with ad neatly removed. **$28.00\£16.00** *(BB 12/88)*

Hamilton, George H. Plants of the Niagara Parks System of Ontario. Toronto, 1943. [xii], 233pp. Gilt stmpd brds, VG. **$55.00\£99.00** *(Hrt 12/89)*

Hampden, Mary. Every Woman's Flower Garden. NY, 1913. 1st Amer ed. 353pp. Ownr's sig, else VG in dec clth. **$35.00\£19.00** *(BF 2/90)*

Hampton, F.A. The Scent of Flowers and Leaves. London, 1925. Brds. a little fxng. Lady Barlow's cpy w/ a few pencil notations. Frnt e.p. tape-marked. **$45.00\£25.00** *(WW 1/90)*

Hanzlik, Edward J. Trees and Forests of Western United States. Portland, OR 1928. 128pp. Gold embossed green cloth. **$16.00\£9.00** *(BB 12/88)*

Another copy. Portland, OR, 1928. Privately printed. 1st ed. 128pp. Some cover spotting else VG. **$35.00\£19.00** *(JI 12/88)*

Harada, Jiro. Japanese Gardens. London, 1956. 160pp. As new in dj. **$25.00\£14.00** *(RW 6/89)*

Harada, Jiro. Japanese Gardens. London, 1956. 160pp. F in dj. **$50.00\£28.00** *(BF 2/90)*

Harada, Jiro. The Gardens of Japan. London, 1928. [viii], 180pp. T.e.g., stmpd clth w/gilt, some shlf wr. **$40.00\£22.00** *(Hrt 12/89)*

Harborne, J. et al. Chemotaxonomy of the Leguminosae. Academic, 1971. 612pp. VG in dj. **$50.00\£28.00** *(GW 8/89)*

Harding, Winton. Saxifrages; The Genus Saxifrage in the Wild and in Cultivation. London, 1970. 134pp. Softbound. **$12.50\£7.00** *(BB 12/88)*

Hardy, George & Winifred. Wild Flowers in the Pacific Northwest. Saskatoon, 1964. 1st ed. 128pp. Inscribed by the publisher. VG in slightly worn dj. **$30.00\£17.00** *(JI 4/89)*

Harler, Agnes, W. The Garden in the Plains. London, 1943. 2nd ed. Rev. & corrected (1941). [xv], 441pp. Printed in India, green buckram covers. G. **$28.00\£16.00** *(BB 12/88)*

Harned, J. Wildflowers of the Alleghanies. Author, 1936. 2nd ed. 675pp. Signed. Cracked front hinge, else VG. **$40.00\£22.00** *(GW 8/89)*

Harrington, H.D. Manual of the Plants of Colorado. Chicago, 1964. 2nd ed (1954). [x], 666pp. with map. Brown cloth, 4to. VG. **$48.00\£27.00** *(BB 12/88)*

Harris, E.C. New Zealand Berries - New Zealand Flowers - New Zealand Ferns. Nelson, 1894. 3 vols. in one. Rare. Binder's clth. A little minor fxng. **$135.00\£75.00** *(WW 1/90)*

Harris, Lena Scott. California Native Flowers. LA, n.d. (1936) Priv. Publ. Spiral bound blue leaterette. Oblong 12mo. 53 heavy card stock pages (of 54 called for). Signed. **$12.50\£7.00 (BB 12/88)**

Harris, Thaddeus William. A Treatise on Some of the Insects Injurious to Vegetation. NY, 1862. New enlrgd ed. 640pp. Tndr, wrn brds, a few stains internally. **$45.00\£25.00** *(Hrt 12/89)*

Harris, Thomas Y. Wild Flowers of Australia. 1958. 4th ed. About F. **$24.00\£13.00** *(PBx 11/89)*

Harrison, C.R. Ornamental Conifers. NY, 1975. 224pp. As New in dj. **$36.00\£20.00** *(Hrt 12/89)*

Harshberger, John. Text-Book of Pastoral and Agricultural Botany. Phila, 1920. 294pp. Spine dull. **$7.50\£4.00** *(WH 1/90)*

Hartmann, Hudson & Dale Kester. Plant Propagation. NJ, 1968. 2nd ed. 702pp. Name f.f.e.p., else VG. **$12.00\£7.00** *(CB 1/90)*

Hartzog, Mattie. Garden Time in the South. McFarland, US, 1951. 246pp. G. **$21.50\£12.00** *(ABx 3/89)*
Another copy. Harrisburg, PA, 1951. [xvi], 246pp. VG in dj. **$10.00\£6.00** *(RW 6/89)*

Harvey, John. Early Gardening Catalogues. London, 1972. 182pp. As new in dj. 8vo. **$21.50\£12.00** *(ABx 3/89)*

Harvey, John. Early Nurserymen. London, 1974. 276pp. As new in dj. **$21.50\£12.00** *(ABx 3/89)*

Harvey, Norman B. New Zealand Botanical Paintings. Christchurch, NZ, 1969. 1st ed. 88pp. Boxed. Folio. Fine. **$85.00\£47.00** *(JI 12/88)*
Another copy. Christchurch, 1969. G in dj. **$54.00\£30.00** *(WW 1/90)*

Harvey-Gibson, R. Outlines of the History of Botany. Black, 1919. 274pp. **$37.50\£21.00** *(GW 8/89)*

Haselton, S. Cacti for Amateurs. Abbey, 1938. 142pp. VG. **$15.00\£8.00** *(GW 12/88)*

Haskin, L. Wildflowers of the Pacific Coast. Metro, 1934. 408pp. VG. **$22.50\£13.00** *(GW 12/88)*

Haskin, Leslie L. Wild Flowers of the Pacific. Portland, 1934. [x], 407pp. Ownr's sig, rbbd else G. **$14.00\£8.00** *(Hrt 12/89)*

Hastings, Somerville. Summer Flowers of the High Alps. London, 1910. 85pp. Blck & glt dec bndnb, some fxng frnt & bck pp. Else VG. **$35.00\£19.00** *(Hrt 12/89)*

Hatcher, H. et al. Johnny Appleseed, A Voice in the Wilderness. NJ, 1953. 76pp. VG. **$24.00\£13.00** *(AB 4/89)*

Hatfield, Audrey W. Pleasures of Herbs. NY, 1965. 1st ed. 186pp. VG in torn dj. **$17.00\£9.00** *(CB 12/89)*

Hatton, Richard G. The Craftsman's Plant-Book: or Figures of Plants. London, 1909. Sm 4to. [ix], 539pp. Own's sig else about F. **$220.00\£122.00** *(Hrt 12/89)*

Hawaiian Sugar Planters' Association. Story of Sugar in Hawaii. Honolulu, 1926. 96pp. Wrps VG. **$10.00\£6.00** *(Hrt 12/89)*

Hawken, Paul. The Magic of Findhorn. NY, 1975. 1st ed. 216pp. VG in dj. **$12.00\£7.00** *(AB 4/89)*

Hawkes, Alex D. Wild Flowers of Jamaica. London, 1947. 1st ed. 96pp. VG. **$7.50\£4.00** *(WH 1/90)*

Hawkes, J.G. (ed). Chemotaxonomy & Serotaxonomy. 1968. 299pp. F. **$24.00\£13.00** *(PBx 11/89)*

Haworth-Booth, Maude. My Garden Diary. London, 1934. 144pp + 10pp of plans + 4pp ads. Spine sunned, else VG. **$20.00\£11.00** *(RW 11/86)*

Haworth-Booth, Michael. Effective Flowering Shrubs. London, 1973. Revised ed. 350pp. G in dj. Garden Book Club. **$12.50\£7.00** *(ABx 3/89)*

Hawthorne, Hildegarde. The Lure of the Garden. NY, 1911. [x], 259pp. T.e.g. ltd stain on gilt dec brds, spine tender else VG. **$75.00\£42.00** *(Hrt 12/89)*
Another copy. NY, 1911. 259pp. Blndstmpd pict green clth, spine barely frayed, darkened. **$80.00\£44.00** *(BF 2/90)*

Hawthorne, Hildegarde. The Lure of the Garden. NY, 1911. [x], 259pp. Small 4to. Decorated cover. VG. **$45.00\£25.00** *(RW 11/86)*

Hay, Thomas. Plants For the Connoisseur. NY, 1938. [xi], 180pp. VG. **$12.00\£7.00** *(RW 11/86)*
Another copy. London, 1938. 180pp. G. **$21.50\£12.00** *(ABx 3/89)*

Hay, Thomas. Plants for the Connoisseur. NY, 1938. 1st Amer ed. 180pp. VG. **$20.00\£11.00** *(BF 2/90)*

Hayata, B. Botanical Survey by the Government of Formosa, with Short Sketches on the Vegetation and Flora of the Island. Brussels, 1911. 60-82 pp. Small 4to. Bound in marbled boards & black cloth by U. Holzer bindery in Boston. E.A. Wilson's copy, inscribed to him by the author. Gold stamped title on the spine. Minimal edge wear. **$38.00\£21.00 (BB 12/88)**

Hayes, G. & Garrison, D. Key to Important Woody Plants of Oregon & Washington. 1960. 227pp. VG. **$25.00\£14.00** *(GW 8/89)*

Hays, Helen A. A Little Maryland Garden. NY, 1909. 201pp. Decorated cover. VG. **$15.00\£8.00** *(RW 11/86)*

Hayward, Ida & George Druce. The Adventive Flora of Tweedside. Arbroath, 1919. 296pp. Library stamp f.f.e.p. else G. **$63.00\£35.00** *(ABx 3/89)*

Hazlehurst, Franklin H. Jacques Boyceau and the French Formal Garden. Athens, 1966. 137pp. VG. **$32.00\£18.00** *(BF 2/90)*

Headrick, J. General View of the Agriculture of the County of Angus or Forfarshire. Edinburgh, 1813. [xxxi], 589pp. Hlf clf (antique style). G. **$108.00\£60.00** *(WW 1/90)*

Healey, B.J. The Plant Hunters. NY, 1975. [vii], 214pp. As New in dj. **$30.00\£17.00** *(Hrt 12/89)*

Healey, B.J. The Plant Hunters. NY, 1975. 214pp. G in torn dj. **$27.00\£15.00** *(ABx 3/89)*

Another copy. NY, 1975. 1st ed. [x], 214pp. VG in dj. **$25.00\£14.00** *(JI 12/88)*

Hearne, D.A. Trees for Darwin and Northern Australia. Canberra, 1975. 143pp. Cr 4to. Brds. G. **$27.00\£15.00** *(WW 1/90)*

Heath, F.G. The Fern Paradise. London, 1878. 5th ed. Trifle used. **$36.00\£20.00** *(WW 1/90)*

Heath, F.G. The Fern Portfolio. London, 1885. 15plts of 45 colored figs., folio, orig clth, neatly rebckd. **$54.00\£30.00** *(WW 1/90)*

Heath, Francis George. The Fern World. London, 1879. 6th ed. 450pp. Decorated cloth. G. **$39.50\£22.00** *(ABx 3/89)*

Hedberg, O. Afroalpine Vascular Plants. Uppsala, 1957. 411pp. Wrpprs, spine trn. G. **$45.00\£25.00** *(WW 1/90)*

Hedla, Lenore. Gardens for Alaskans. n.p., 1974. 206pp. Wrps. VG. **$5.00\£3.00** *(WH 1/90)*

Hedrick, U.P. A History of Agriculture in the State of New York. NY, 1966. [xxxx], 465pp. Pbck. **$12.00\£7.00** *(WW 1/90)*

Hedrick, U.P. A History of Horticulture in America to 1860. NY, 1950. 551pp. F in dj. **$100.00\£56.00** *(BF 2/90)*

Hedrick, U.P. et al. The Peaches of New York. Report of the New York Agricultural Experiment Station, 1916, 1917. [xiii], 541pp. Rebound in green buckram 4to. G. **$135.00\£75.00** *(ABx 3/89)*

Hedrick, U.P. et al. The Plums of New York. Report of the New York Agricultural Experiment Station, 1910, 1911. [xii], 616pp. Rebound in green buckram 4to. G. **$135.00\£75.00** *(ABx 3/89)*

Hedrick, U.P. Grapes and Wines from Home Vineyards. NY, Oxford, 1945. 326pp. VG in wrn, trn dj. **$18.00\£10.00** *(CB 1/90)*

Hedrick, U.P. Manual of American Grape-Growing. NY, 1924. New & rev. [xiii], 458pp. G. **$30.00\£17.00** *(Hrt 12/89)*

Hegi, Dr. Gustave von. Illustrierte Flora von Mittel-Europa. 2nd ed. 1937-38. 7 vols in 13. Orig blue clth, dec cvrs. Scarce. Nice clean set. **$1,170.00\£650.00** *(DL 89/90)*

Heinrich, Julius J. The Window Flower Garden. NY, 1897. New & enlrgd ed. [vi], 123pp. Orig green stmpd, gilt clth. Lght spttng to cvr, else brght, VG. **$32.00\£18.00** *(CB 1/90)*

Heiser, Charles. Nightshades, Pardoxical Plants. SF, 1969. 200pp. Nearly new in dj. **$15.00\£8.00** *(AB 4/89)*

Heline, Corinne. Magic Gardens; Symbolic Rendering of Angelic Communion with Man through the Medium of Flowers. LA, 1944. 122pp. Faded spine and paper covered boards. **$25.00\£14.00** *(AB 4/89)*

Hellyer, A.G.L. Amateur Gardening Picture Book of Gardens. London, 1956. 72 full-pg hlf tones. Head/foot of spine faded else VG in wrn dj. **$15.00\£8.00** *(CB 1/90)*

Hellyer, A.G.L. Shrubs in Color. London, 1965. 3rd imp. 128pp. Few watr stns to frnt covr; bottm edge faded else VG in reprd dj. **$13.00\£7.00** *(CB 1/90)*

Hellyer, AGL & Lloyd, Christoper. Amateur Gardening Photo Album of Garden Plants. London, 1961. 1st ed. 192pp. Oblong 8vo. VG/F in dj. **$11.50/£6.00** *(BLB 7/89)*

Hellyer, AGL & Lloyd, Christopher. Encyclopedia of Plant Portraits. London, 1953. 1st ed. 322pp. VG/F in worn dj. **$10.00\£6.00** *(BLB 7/89)*

Hemenway, H.D. How to Make Home and City Beautiful. Northampton, MA, 1911. 106pp. Top of covers water stained. **$15.00\£8.00** *(RW 11/86)*

Henderson, G. & Addison, M. Malayan Orchid Hybrids. Singapore, 1963. 187pp. Popular ed., VG in dj. **$20.00\£11.00** *(GW 8/89)*

Henderson, J. Henderson's Hand-Book of the Grasses of Great Britain and America. Journal, 1875. 238pp. Worn. Hinge cracked. **$16.00\£9.00** *(GW 12/88)*

Henderson, M.R. Malayan Wild Flowers, 1954-1959. 2 vols. 1974. Nice set in djs. **$72.00\£40.00** *(DL 89/90)*

Henderson, Peter. Garden & Farm Topics. NY, 1884. 244pp. Cloth, worn, a bit shaken. **$22.00\£12.00** *(AB 2/89)*

Henderson, Peter. Gardening for Pleasure. NY, 1902. Green clth, enlrgd ed. 404pp. Shelf worn, else VG. **$10.00\£6.00** *(11/88)*

Henderson, Peter. Gardening for Profit... NY, 1874. New and Enlarged edition. 276pp. VG $5.00\£3.00 *(RW 11/86)*

Another copy. NY, 1886. Green cloth, 276 pp. VG. **$10.00\£6.00 (MD 11/88)**

Another copy. NY, 1894. 376pp. As new. **$20.00\£11.00** *(AB 2/89)*

Henderson, Peter. Henderson's Handbook of Plants. NY, 1910. 2nd ed. 528pp. Large 8vo. G. **$21.00\£12.00** *(CB 8/89)*

Henderson, Peter. Practical Floriculture. NY, 1874. 288pp. Brown clth, enlrgd. ed. Minor waterstain effects end papers and front corner, else VG. **$10.00\£6.00** *(MD 11/88)*

Hennedy, Robert. The Clydesdale Flora. Glasgow, 1878. In memoriam ed revised. 256pp. Two gatherings loose. G. **$32.50\£18.00** *(ABx 3/89)*

Another copy. Hopkins, 1891. 277pp. VG. **$15.00\£8.00** *(GW 8/89)*

Hennessey, Roy. Hennessey on Roses. Portland, OR, 1943. 2nd ed. 191pp. VG. **$7.00\£4.00** *(CB 12/89)*

Henrey, Blanche. Flower Portraits. London, 1937. 127pp. G. **$14.50\£8.00** *(ABx 3/89)*

Henry, Joseph Kaye. Flora of Southern British Columbia and Vancouver Island. Toronto, 1915. 1st ed. [xvi], 368pp. Label f.e.p. else VG. **$45.00\£25.00** *(JI 4/89)*

Henshaw, Julia. Mountain Wild Flowers of America. Boston, 1906. 1st ed. [xxii], 386pp. Slight cov wear else VG. **$55.00\£31.00** *(JI 4/89)*

Henshaw, Julia. Wildflowers of the North American Mountains. NY, 1917. 2nd ed. 383pp. Inscribed by author. VG. **$25.00\£14.00** *(RW 11/86)*

Henslow, J.S. and E. Skepper. Flora of Suffolk. London, 1860. [x], 140pp. Orig. clth. title-pge trifle fxed and endpapers discolored. **$45.00\£25.00** *(WW 1/90)*

Henslow, T. Geoffrey. Garden Construction. London, 1923. 285pp. G. **$39.50\£22.00** *(ABx 3/89)*

Henslow, T. Geoffrey. Ye Sundial Booke. London, 1914. 422pp. Gilt deocorated cream boards, spine yellowed. VG. Presentation copy, signed. **$198.00\£110.00** *(ABx 3/89)*

Henslow, T.G.W. The Rose Encyclopedia. London, 1934. 320pp. New edn., revised by Walter Brett. Becoming scarce. G/VG. **$11.50\£6.00** *(BLB 7/89)*

Hepting, G.H. Diseases of Forest and Shade Trees of the United States. USDA hndbk #386, 1971. 658pp. Unused. **$20.00\£11.00** *(PBx 11/89)*

Herbarum Imagines Vivae. 1535 (1966). Ltd. ed. #22 of 300. Marblized paper covers, velour lined, sheepskin covered box. Rare. Immaculate. **$75.00\£42.00** *(PBx 11/89)*

Hereman, S. Blight on Flowers. London, 1840. [xi], 242pp. New clth, plts fxed, stain on frontispiece. **$86.00\£48.00** *(WW 1/90)*

Hersey, Jean. I Like Gardening. Boston, 1941. 1st ed. 234pp. Quite G in dj. **$16.00\£9.00** *(Hrt 12/89)*

Hersey, Jean. The Shape of a Year. NY, 1967. [x], 243pp. VG in dj. **$15.00\£8.00** *(Hrt 12/89)*

Hertrich, W. Camellias in the Huntington Gardens. Huntington, 1955. 3 vols. 1124pp. Slightly worn djs. All 3 signed presentation copies. VG. **$100.00\£56.00** *(GW 12/88)*

Hertrich, William. Camellias in the Huntington Gardens. San Marino, 3 vols. Vol. I: 1954, [ix, i], 380pp, pres cpy dated Sept. 21, 1955. Vol. II: 1955, [vii, i], 378pp. Vol. III: 1959, [xii], 366pp. VG in dj. **$165.00\£92.00** *(Hrt 12/89)*

Hertrich, William. Palm and Cycads. San Marino, 1951. 1st ed., [x], 141pp with color frontis. VG. **$37.50\£21.00** *(BB 12/88)*

Hertrich, William. The Huntington Botanical Gardens, 1905-1949. San Marino, 1949 - 1st ed. (1953 - 2nd prtng.) 4to. 167pp. Decorated boards, cloth spine with paper label. Slight corner wear, else VG. **$26.00\£14.00** *(BB 12/88)*

Another copy. 167pp. 1949. Signed pres copy. **$26.00\£14.00** *(GW 12/88)*

Hervey, A.B. Sea Mosses. Boston, 1882. 20 monochrome plts. Sound, crisp and clean. **$45.00\£25.00** *(PBx 11/89)*

Hesler, L. North American Species of Hygrophorus. TN, 1963. 416pp. VG **$15.00\£8.00** *(GW 8/89)*

Another copy. Knoxville, 1963. 416pp. F. **$12.50\£7.00** *(WH 1/90)*

Hibberd, S. New & Rare Beautiful-Leaved Plants. Bell, 1870. 144pp. Top of spine chipped. **$65.00\£36.00** *(GW 12/88)*

Hibberd, Shirley. Amateur's Greenhouse and Conservatory. London, nd(1890's) rev. ed. 301pp. G. **$63.00\£35.00** *(ABx 3/89)*

Hibberd, Shirley. The Fern Garden. London, 1869. 2nd ed. 148pp + ads. Small split at top of back hinge, else a nice, bright copy. **$86.50\£48.00** *(ABx 3/89)*

Hickok, Peggy. Home Gardens in Hawaii. Honolulu, 1958. 128pp. 4to. VG $24.00\£13.00 *(BB 12/88)*

Hicks, Sam. Desert Plants and People. San Antonio, 1966. [xvi], 75pp. Inscribed by author. Near F, in dj. **$24.00\£13.00** *(BB 12/88)*

Higgins, V. Crassulas in Cultivation. London, 1964. G. **$21.00\£12.00** *(WW 1/90)*

Higgins, Vera. The Study of Cacti. London, 1950. Reprint. 144pp. G. **$7.25\£4.00** *(ABx 3/89)*

Hill, Amelia Leavitt. Garden Portraits. NY, 1923. 280pp. Few pp roughly opened, ownr's inscrip on hlf-title else VG in slightly rbbd pict clth. **$20.00\£11.00** *(BF 2/90)*

Hill, Amelia. Garden Portraits. NY, 1923. 1st ed. 230pp. Spine dull, inscription laid in. G/VG $8.00\£4.00 *(BLB 7/89)*

Hill, Anna Gilman. Forty Years of Gardening. NY, 1938. 301pp. Sgnd on title pge. Ownr's bkplte; spine slightly sunned and soiled. **$25.00\£14.00** *(BF 2/90)*

Hill, Clara Chapman. Spring Flowers of the Lower Columbia Valley. Seattle, 1958. [xi], 164pp. Blue marbled board. Fine in dj. **$19.50\£11.00** *(BB 12/88)*

Hill, J. The British Herbal. London, 1756. pp. [iv], 533, [3]. Folio, mdrn hlf clf. Some slght waterstning, last leaf shows use else G. **$864.00\£480.00** *(WW 1/90)*

Hill, J. The Family Herbal. Bungay, 1812. [viii, xl], 376pp. Mdrn hlf clf, antique style. As usual shows some slght signs of use. **$216.00\£120.00** *(WW 1/90)*

Hill, Jason. The Curious Gardener. London, 1932. 173pp. E.p.s. slgltly fxd else VG. **$30.00\£17.00** *(BF 2/90)*

Hills, L.D. The Propagation of Alpines. London, 1950. 464pp. G in dj. **$54.00\£30.00** *(WW 1/90)*

Hills, Lawrence D. Alpines Without a Garden. London, 1953. 1st ed. 192pp. G in torn dj. **$18.00\£10.00** *(ABx 3/89)*

Hills, Lawrence D. The Propogation of Alpines. London, 1950. 464pp. VG in dj. **$20.00\£11.00** *(RW 11/86)*
Another copy. London, 1959. 2nd ed. 484pp. G in torn dj. **$72.00\£40.00** *(ABx 3/89)*

Hinton, P. Eden Phillpotts, A Bibliography of First Editions. Birmingham, 1931. Hf vellum. Portrait signed by Phillposst. #79 of 350. 164pp. Title page signed by Hinton. G. **$135.00\£75.00** *(PBx 11/89)*

Hitchcock, A. A Flora of California: Gramineae. CA, 1912. 107pp. Wrs. VG. **$15.00\£8.00** *(GW 12/88)*

Hitchcock, A. Genera of the Grasses of the United States. USDA, 1920. 307pp. Bound. A few pencil notes, else VG. **$25.00\£14.00** *(GW 8/89)*

Hitchcock, A. The Grasses of Central America. USNH, 1930. 205pp. Wrs. VG. **$15.00\£8.00** *(GW 12/88)*

Hitchcock, A. The Grasses of Ecuador, Peru & Bolivia. USNH, 1927. 265pp. Wrs. VG. **$15.00\£8.00** *(GW 12/88)*

Hitchcock, A.S. Manual of the Grasses of the U.S. Wash, 1935. 1040pp. Binding dull. **$20.00\£11.00** *(WH 1/90)*

Hitchcock, A.S. Manual of the Grasses of the West Indies. 1936. 439pp. Unused. **$24.00\£13.00** *(PBx 11/89)*

Hjelt, H. Conspectus Florae Fenniae. London, 1888-1923. Vols. 1-6. In 3 vols. Hlf clth. G. **$135.00\£75.00** *(WW 1/90)*

Hoare, A.H. The English Grass Orchard. London, 1928. 227pp. Trifle wrn. **$21.00\£12.00** *(WW 1/90)*

Hoare, A.H. The English Grass Orchard. London, 1928. 227pp. VG. **$35.00\£19.00** *(RW 6/89)*

Hoare, Henry. Calendar of Flowering Trees and Shrubs. London, n.d. (1920's). [xxxii], 117pp. One page torn & repaired, else G. **$29.00\£16.00** *(ABx 3/89)*

Hobson, N.K., J.P. Jessop, M.C. Ginn & J. Kelly. Veld Plants of South Africa. 1975. As New in dj. **$72.00\£40.00** *(DL 89/90)*

Hodgson, N. Barrie. Grasses, Sedges, Rushes and Ferns of the British Isles. London, 1949. 93pp. VG in dj. **$15.00\£8.00** *(RW 6/89)*

Hoehne, F.C. (ed). Flora Brasilica, Papilionaceae, Aristolochiaceae, Labiatae, Onagraceae, Lycopodiaceae and Xyridaceae. San Paulo, 1948-1968. 4to. Orig wrpprs. G. **$216.00\£120.00** *(WW 1/90)*

Hogg, Warrington. The Book of Old Sundials & Their Mottoes. Edinburgh, 1922. New imp. 103pp. Sm portion of head of spine mssng, slght smudge/dent on paper dec brds else F. **$65.00\£36.00** *(Hrt 12/89)*

Hohnel, F.V. Fragmente Zur Mykologie. Vienna, 1967. 2032pp. 2 vols. G. **$81.00\£45.00** *(WW 1/90)*

Holden, Alexander E. Plant Life in the Scottish Highlands. London, 1952. 1st ed. 319pp. G in dj. **$43.00\£24.00** *(ABx 3/89)*
Another copy. London, 1952. [xv], 319pp. Clth. G. **$36.00\£20.00** *(WW 1/90)*

Holdgate, M. Antarctic Ecology. Academic, 1970. Vol. 2. 391pp. VG. **$15.00\£8.00** *(GW 8/89)*

Hole, (Dean) S. Reynolds. A Book About Roses. London, 1904. 21st ed. 300pp. Gilt dec brds, sl tender, shlf wr else VG. **$18.00\£10.00** *(Hrt 12/89)*

Hole, S. Reynolds. Our Gardens. London, 1899. 1st ed. [x], 304pp. Endpapers and some text drawings by Rackham. t.e.g. VG. **$35.00\£19.00** *(RW 11/86)*
Another copy. London, 1901. 4th ed. 302pp. G. **$36.00\£20.00** *(ABx 3/89)*
Another copy. London, 1899. 2nd ed. 304pp. Cloth rubbed and corners bumped. **$10.00\£6.00** *(AB 2/89)*

Holland, W. The Moth Book. Doubleday, 1922. 479pp. A few nicks, else VG. **$20.00\£11.00** *(GW 8/89)*

Holley, W.D. & Ralph Baker. Carnation Production. N.p., (Pub. by Colorado Carnations), 1963. [ix], 142pp. With b&w illus. 4to, gilt decorated cloth. VG. **$19.00\£11.00** *(BB 12/88)*

Hollingsworth, Buckner. Flower Chronicles. NJ, 1958. 302pp. VG $15.00\£8.00 *(RW 6/89)*

Hollingsworth, Buckner. Gardening on Main Street. NJ, 1968. 170pp. VG in dj. **$5.00\£3.00** *(RW 11/86)*

Hollingsworth, Buckner. Her Garden Was Her Delight. NY, 1962. 166pp. VG in dj. **$25.00\£14.00** *(BF 2/90)*

Holly, Henry H. Holly's Country Seats. NY, 1863. 171pp. Green cloth, spine dull, joints broken, some signatures sprung, severe shelf wear, plates have minor foxing. **$165.00\£92.00** *(MD 11/88)*

Holme, Charles (ed.) The Gardens of England, in the Northern Counties. London, 1911. 136 photoplts, 8 col. [xxxvi]. Orig wrps VG. **$65.00\£36.00** *(Hrt 12/89)*

Holme, Charles (ed.) The Gardens of England in the Midland & Eastern Counties. London, 1908. 136 col/bw plts, [x, xxxviii]. G. **$60.00\£33.00** *(Hrt 12/89)*

Holme, Charles (ed). The Gardens of England in the Midland and Eastern Counties. London, The Studio, 1908. 4to. [x], [xxvii]pp. 1/2 leather, marbled endpapers. VG. **$65.00\£36.00** *(RW 11/86)*

Holmgren, Arthur H. & Berniece A. Andersen. Handbook of the Vascular Plants of the Northern Wasatch. N.p., 1959. 2nd ed (1947). [v], 202pp. Spiral bound blue leatherette paper. Well used but intact. **$12.00\£7.00** *(BB 12/88)*

Holtz, Mathilde & Katherine Bemis. Glacier National Park, Its Trails and Treasures. NY, 1917. [xv], (4), 19-263pp. Photo/glt pict brds. G. **$20.00\£11.00** *(Hrt 12/89)*

Honig, Peter & Frans Veerdoorn, eds. Science & Scientists in the Netherlands Indies. NY, 1945. [xxii], 491pp. Blnd stmp few pp, rbbed else quite G. **$50.00\£28.00** *(Hrt 12/89)*

Hooker, W.J. & J.G. Baker. Synopsis Filicum. London, 1874. 2nd ed. [xiv], 9-559pp. 1-8 were not publshd. New clth. **$126.00\£70.00** *(WW 1/90)*

Hooker, W.J. Niger Flora, or an Enumeration of the Plants of Western Tropical Africa. London, 1849. [xvi], 587pp. Trifle used, stmp on title-page and bck of plts. **$180.00\£100.00** *(WW 1/90)*

Hooker, W.J. Species Filicum. London, 1846-64. 3 vols., old hlf lthr, 2vols. Orig clth. Lib stmps on titles & rev of plates. Vols 3 & 4 neatly repred. Almost unobtainable orig prntng. **$396.00\£220.00** *(WW 1/90)*

Hooker, W.J. The Journal of Botany. London, 1834-42. 4 vols. vol 4 has title pge, 55-56pp, 169-198pp, 393-436pp. All publshd and very rare. Bndngs a little wrn and soiled. Library stmps on title-pp and reverse of plates. **$324.00\£180.00** *(WW 1/90)*

Hooker, William Jackson. Flora Scotica. Edinburgh & London, 1821. 2 vols. [x], 292;297. Half calf with marbled boards, generally worn but hinges sound. **$162.00\£90.00** *(ABx 3/89)*

Hooker, William Jackson. Journal of a Tour in Iceland in the Summer of 1809. London, 1813. 2nd ed. 2 vols. Vol I: [cvi], 369pp. Vol. II: 391pp. (15) of index. Early hlf lthr on mrbld brds, corners wrn. Pp lght uniform brwning in nice tight cpy, institutional blnd stmp on several plts & pp else VG. **$380.00\£211.00** *(Hrt 12/89)*

Hope, C.W. The Ferns of North-Western India. Bombay, 1899-1904. 224pp. Hlf clf (somewhat wrn) $63.00\£35.00 *(WW 1/90)*

Hopkirk, T. Flora Glottiana. Glasgow, 1813. [ii], iv, [ii], 6-170pp. New clth, stmp on title and hlf-title. G. **$108.00\£60.00** *(WW 1/90)*

Hornibrook, M. Dwarf and Slow-Growing Conifers. London, 1973. 2nd ed. [xvi], 286pp. **$21.00\£12.00** *(WW 1/90)*

Hort, Sir Arthur. Garden Variety. London, 1935. 255pp. Slght fxng, prt of rear e.p. lacking. Ownr's inscrip. Slightly wrn clth. **$30.00\£17.00** *(BF 2/90)*

Hort, Sir Arthur. The Unconventional Garden. London, 1928. 280pp. G. **$23.50\£13.00** *(ABx 3/89)*

Hottes, Alfred C. Garden Facts and Fancies. NY, 1949. 370pp. VG in dj. **$35.00\£19.00** *(Hrt 12/89)*

Hough, R.B. Handbook of the Trees of the Northern States and Canada. NY, 1924. 3rd rev. ed. (1907). [x], 470pp. Minimal cover soiling else VG. **$44.00\£24.00** *(BB 12/88)*

House and Garden. Bound vol. July-December, 1915. Gray cloth, somewhat wrn. About VG. **$35.00\£19.00** *(AT 3/89)*

House, H.D. Wild Flowers. NY, (1934) 1936. 362pp. G. **$48.00\£27.00** *(WW 1/90)*

House, Homer D. Bibliography of the Botany of New York State 1751-1940. NY State Museum Bull. Albany, 1942. 2 vols. Prt 1, 174pp. Prt 2, 175-233pp. Wrps VG. **$32.00\£17.00** *(Hrt 12/89)*

House, Homer D. Wild Flowers. NY, 1935. 362pp. Blck stmpd clth. VG. **$70.00\£39.00** *(Hrt 12/89)*

Howard, Alexander L. Trees in Britain and Their Timbers. London, 1949. [xii], 340pp. VG in worn dj. **$7.50\£4.00** *(RW 11/86)*

Howard, C.W. & K.P. Buswell. A Survey of the Silk Industry of South China. Hong Kong, 1925. 208pp. Wrpprs. G. **$36.00\£20.00** *(WW 1/90)*

Howard, Edwin L. Chinese Garden Architecture. NY, 1931. 50 b/w photos, x, brief text preceding each plt. brd spotted. Spine dark & tender, some red underlining in text. **$35.00\£19.00** *(Hrt 12/89)*

Howard, Sir Albert. An Agricultural Testament. NY, 1949. 5th prtg. [xv], 253pp. Near F. **$15.00\£8.00** *(Hrt 12/89)*

Howe, Bea. Lady with Green Fingers. London, 1961. 184pp. G in dj. **$25.00\£14.00** *(ABx 3/89)*

Howe, M.A. DeWolfe. Boston Common, Scenes from Four Centuries. Boston, 1921. 89pp. VG in sun stained brds. **$45.00\£25.00** *(Hrt 12/89)*

Howells, Victor. A Naturalist in Palestine. NY, 1957. 180pp. VG in older dj. **$30.00\£17.00** *(Hrt 12/89)*

Hoyt, Roland Stewart. (compiled & arrngd) Check List for the Ornamental Plants of Sub-Tropical Regions. Los Angeles, 1938. [x], 383pp. Spine fade, sm puncture in spine, else VG. **$18.00\£10.00** *(Hrt 12/89)*

Hubbard, Henry & Theodora Kimball. An Introduction to the Study of Landscape Design. NY, 1935. Rev. ed, 419pp. VG. **$35.00\£19.00** *(RW 11/86)* *Another copy.* NY, 1927. 406pp. Green clth, spine dull, shlf wrn, some fxng else VG. **$32.50\£18.00** *(MD 2/90)*

Hubbard, Henry V. and Theodora Kimball. An Introduction to the Study of Landscape Design. NY, 1924. 2nd prntng. 406pp. Near F in slightly rbbd clth. **$60.00\£33.00** *(BF 2/90)*

Hudson, Derek. Kensington Palace. London, 1968. [xvii], 141pp. VG in dusty dj. **$36.00\£20.00** *(Hrt 12/89)*

(Hudson, W.H.) 153 Letters from W.H. Hudson, edited and with an introduction and explanatory notes by Edward Garnett. London, 1932. Nonesuch Press, ltd ed. 372/1000. 191pp. Near F. **$120.00\£67.00** *(Hrt 12/89)*

Hudson, W.H. The Naturalist in La Plata. London, 1903. 4th ed. [ix], 394pp. T.e.g. sl tendr in dec brds, gilt lettering spine, else VG. **$35.00\£19.00** *(Hrt 12/89)*

Hudsoni Gulielmi. Flora Anglica; Exhibens Plantas per Regnum Britannie. 1778. 690pp. 2 vols bnd in 1. Fll clf. Well preservd cpy. **$144.00\£80.00** *(DL 89/90)*

Hui-Lin, Li. The Origin and Cultivation of Shade and Ornamental Trees. Phil, 1963. 282pp. VG in soiled, chppd dj. **$20.00\£11.00** *(CB 1/90)*

Hull, George F. Bonsai for Americans. NY, 1964. 1st ed. 256pp. G. **$18.00\£10.00.**

Hull, George F. The Language of Gardening. World, 1967. 191pp. G in dj. **$9.00\£5.00** *(ABx 3/89)*

Hulme, F. Edward. Familiar Wild Flowers. London, et al, 1912. 160pp. Small 8vo. G. **$10.00\£6.00** *(CB 12/89)*

Hulme, F.E. & Hibberd, S. Familiar Garden Flowers. London, nd (1870's). 5 vols bound in 2. Vol I, [xvi], 160, [xvi], 160, [xvi], 160. Vol II, [xvi], 160, [xv], 167. 8vo half calf. G. **$126.00\£70.00** *(ABx 3/89)*

Humboldt, Alexandre de & Aime Bonpland. Essai sur la Geographie des Plantes. Paris, 1805. Mexico D.F., 1955. Ltd ed., thin folio, #232/510. [xii], 155pp. About F in dj. **$75.00\£42.00** *(Hrt 12/89)*

Hume, H. Gardening in the Lower South. Mcmillan, 1929. 453pp. VG. **$12.00\£7.00** *(GW 8/89)*

Hume, H. Harold. Camellias. NY, 1951. 271pp. G. **$18.00\£10.00** *(ABx 3/89)*

Hume, H. Harold. Gardening in the Lower South. NY, 1968. [xii], 377pp. F in chipped dj. **$25.00\£14.00** *(AT 3/89)*

Hume, Harold. Camellieas in America. Harrisburg, 1946. 350pp. VG in worn, chipped dj. **$40.00\£22.00** *(GW 8/89)*

Humphris, Ted. Garden Glory. London, 1969. 212pp. G in dj. **$14.50\£8.00** **(ABx 3/89)**

Hunn, C.E. & L.H. Bailey. Practical Garden-Book. NY, 1909. 7th ed. 250pp. Cloth soiled, contents VG. **$10.00\£6.00** *(AB 2/89)*

(Hunt) The Rachel McMasters Miller Hunt Botanical Library. Pittsburgh, 1961. 35pp. VG. **$18.00\£10.00** *(Hrt 12/89)*

Hunt, Charles B. Plant Ecology of Death Valley, California. Washington, 1966. [iv], 68pp. **$28.00\£16.00** *(BB 12/88)*

Hunt, John Dixon and Peter Willis. The Genius of the Place—the English Landscape Garden 1620-1820. NY, 1975. 1st Amer ed. 390pp. F in dj. **$50.00\£28.00** *(BF 2/90)*

Hunt, Peter. The Shell Garden's Book. London, 1964. 319pp. VG. **$20.00\£11.00** *(RW 6/89)*

Hunt, Rachel McMasters Miller. Catalogue of Botanical Books in the Collection of R.M.M. Hunt. Pittsburgh. Ltd. ed. of 750. 2 vols in 3. Vol. I: 1958, [lxxiv], 517pp. Vol. II, prt 1, 1961, [ccxliv]; Vol II, Prt 2, 1961, [ix], 655pp. Orig gilt stmpd green clth. Handmade paper. As New. **$1400.00\£778.00** *(Hrt 12/89)*

Hunt, Rachel. William Penn Horticulturist. Pitts, 1953. 38pp. Ltd. ed. inscribed by author. VG. **$15.00\£8.00** *(RW 11/86)*

Hunt, Thomas F. Cereals in America. NY, 1915. 421pp. Hinges reinforced, cloth and contents VG. Pencil notes and marginalia. **$18.00\£10.00** *(AB 4/89)*

Hussey, C. English Gardens and Landscapes 1700-1750. London, 1967. 174pp. G. **$36.00\£20.00** *(WW 1/90)*
Another copy. London, 1967. 174pp. F in dj. **$165.00\£92.00** *(BF 2/90)*

Hutchens, Alma R. Indian Herbalogy of North America. Windsor, Ont., 1973. 3rd ed. [xxxviii], 381pp. Owner's sig else VG. **$30.00\£17.00** *(Hrt 12/89)*

Hutcheson, Martha Brookes. The Spirit of the Garden. Boston, 1923. 221pp. Spine fade else VG. **$44.00\£24.00** *(Hrt 12/89)*
Another copy. Boston, 1923. 221pp. Few pp fxd else VG. **$35.00\£19.00** *(BF 2/90)*

Hutcheson, Martha Brookes. The Spirit of the Garden. Boston, 1923. 221pp. 4to. G. **$81.00\£45.00** *(ABx 3/89)*
Another copy. Boston, 1923. 4to. 221pp. Blue cloth, spine dull minor wear. **$30.00\£17.00** *(MD 7/89)*

Hutchinson, John. A Botanist in South Africa. London, 1946. [xii], 686pp. Slightly tendr. Scarce. **$90.00\£50.00** *(Hrt 12/80)*
Another copy. 1946. 1st and only ed. Scarce. Nice copy. **$72.00\£40.00** *(DL 89/90)*

Hutchinson, John. A Botanist in Southern Africa. London, 1946. [xii], 686pp. Very scarce. VG. **$75.00\£42.00** *(WH 1/90)*

Hutchson, J. & J.M. Dalziel. Flora of West Tropical Africa. London, various dates, 1954-72. 2nd ed. Rev. 3 vols in 5. 1945pp. VG set in printed wraps. **$85.00\£47.00** *(BB 12/88)*

Huxley, A. Mountain Flowers. London, 1967. 428pp. G. **$32.00\£18.00** *(WW 1/90)*

Huxley, Anthony. Plant and Planet. London, 1974. 428pp. G in dj. **$21.50\£12.00** *(ABx 3/89)*

Hyams, E. & E. Smith. English Cottage Gardens. London, 1970. 250pp. Orig ed. F in dj. **$36.00\£20.00** *(WW 1/90)*

Hyams, E. The English Garden. London, 1964. 288pp. G. **$54.00\£30.00** *(WW 1/90)*
Another copy. London, 1964. 288pp. Folio. VG in slightly wrn dj. **$75.00\£42.00** *(BF 2/90)*
Another copy. London, 1964. 288pp. Scarce. Pg edges brwning else VG in brght, reprd dj. **$75.00\£42.00** *(CB 1/90)*

Hyams, Edward & A.A. Jackson (eds). The Orchard and Fruit Garden. London, 1961. [xv], 208pp. Gilt dec green cloth, top edge gilt. F in dj, red slip case. **$88.00\£49.00** *(BB 12/88)*

Hyams, Edward & Edwin Smith. The English Garden. Ny, 1966. 288pp. Wraps. VG. **$15.00\£8.00** *(RW 6/89)*

Hyams, Edward & William MacQuitty. Great Botanical Gardens of the World. NY, 1969. 1st Amer ed. 288pp. Folio. F in dj. **$65.00\£36.00** *(BF 2/90)*

Hyams, Edward & William MacQuitty. Irish Gardens. London, 1967. 160pp. F in chppd and taped dj. **$65.00\£36.00** *(BF 2/90)*
Another copy. London, 1967. 160pp. VG in worn, reprd dj. **$45.00\£25.00** *(CB 1/90)*

Hyams, Edward. A History of Gardens and Gardening. London, 1971. 345pp. Sm stain on title pg else F in slightly nicked dj and slipcase which is starting to split. **$70.00\£39.00** *(BF 2/90)*
Another copy. NY, 1971. [ix], 345pp. F in brght, chppd dj. Plain box, soiled. **$60.00\£33.00** *(CB 1/90)*

Hyams, Edward. A History of Gardens and Gardening. NY, 1971. [ix], 345pp. Scarce. VG. **$49.00\£27.00** *(BB 12/88)*

Hyams, Edward. An Englishman's Garden. London, 1967. 160pp. VG in slightly rbbd dj. **$35.00\£19.00** *(BF 2/90)*

Hyams, Edward. Capability Brown and Humphry Repton. London, 1971. 248pp. VG in dj. **$35.00\£19.00** *(BF 2/90)*

Hyams, Edward. Capability Brown and Humphry Repton. NY, 1971. 248pp. VG in dj. **$36.00\£20.00** *(ABx 3/89)*

Hyams, Edward. Gardener's Bedside Book. London, 1968. 286pp. As new in dj. **$12.00\£7.00** *(AB 2/89)*

Hyams, Edward. Irish Gardens. NY, 1967. Square 4to. Blue cloth. 160pp. Some minor pencilled marginalia, edges of dj torn. **$50.00\£28.00** *(MD 7/89)*
Another copy. NY, 1967. Lge 4to. 160pp. VG in dj. **$45.00\£25.00** *(RW 6/89)*

Hyams, Edward. Odhams Fruit Grower's Encyclopedia. London, 1960. 253pp. G in dj. **$17.50\£10.00** *(BB 12/88)*

Hyams, Edward. Ornamental Shrubs for Temperate Zone Gardens. NY, 1965. 319pp. Green clth, VG in trn dj. **$20.00\£11.00** *(MD 2/90)*

Hyams, Edward. Pleasure from Plants. London, 1966. 222pp. VG. **$25.00\£14.00** *(BF 2/90)*

Hyams, Edward. The English Garden. London, 1974. New ed of 1964 original. Folio. VG in dj. **$45.00\£25.00** *(RW 6/89)*
Another copy. NY, 1964. 4to. 288pp. VG in dj. **$95.00\£52.00** *(RW 6/89)*

I

Illott, Charles. Book of Asparagus. London, 1901. 108pp. VG. **$20.00\£11.00** *(AB 2/89)*

Ingwersen, W. Classic Garden Plants. London, 1975. 192pp. Sm folio. VG in dj. **$25.00\£14.00** *(BF 2/90)*

Ingwersen, W. The Dianthus. London, 1949. G. **$27.00\£15.00** *(WW 1/90)*

Ingwersen, Will. Classic Garden Plants. London, 1975. 192pp. G in dj. **$36.00\£20.00** *(ABx 3/89)*

Inn, H. Tropical Bloom. Waikiki, 1944. Sm. folio. Portfolio of 40 plates. Lacks outer label else VG. **$20.00\£11.00** *(PBx 11/89)*

Inn, Henry & S.C. Lee. Chinese Houses and Gardens. NY, Bonanza, 1950. 4to. Orange boards, VG in worn dj. **$35.00\£19.00** *(MD 11/88)*
Another copy. NY, 1950. [xii], 148pp. VG in dj. **$50.00\£28.00** *(RW 6/89)*

Innes, William T. Goldfish Varieties and Water Gardens. Phil, 1949. 3rd ed. 385pp. Scarce. G. **$36.00\£20.00** *(Hrt 12/89)*

Innes, William T. Goldfish Varieties and Water Gardens. Phila, 1949. 2nd ed. 386pp. VG. **$25.00\£14.00** *(RW 6/89)*

International Botanical Congress, Proceedings, Abstracts Papers of the XI. Wash, DC & Seattle, 1970. 2 vols. 8vo & 4to, limp brds. VG. **$27.50\£15.00** *(WH 1/90)*

International Botanical Congress, Proceedings of the IX. Montreal, 1959. Vol 2. 444pp. Orig. wrappers. F. **$17.50\£10.00** *(WH 1/90)*

International Grassland Congress, Proceedings of the Sixth. Penn State College, 1952. 2 vols. 1801pp. Green cloth, gilt. F. **$50.00\£28.00** *(WH 1/90)*

Irving, Walter & Reginald Malby. Saxifrages. London, nd. [xii], 147pp. VG. **$12.50\£7.00** *(RW 11/86)*

Ishimoto, Tatsuo. Art of Growing Miniature Trees, Plants and Landscapes. NY, 1956. 143pp. Cloth in dj, contents with light wear. **$6.00\£3.00** *(AB 4/89)*

Ishimoto, Tatsuo. Art of the Japanese Garden. NY, 1960. 3rd prtng. 128pp. VG in lightly worn dj. **$10.00\£6.00** *(AB 4/89)*

J

Jackson, B. Daydon. A Glossary of Botanic Terms. London/NY, 1928. 4th ed. rev. 481pp. Later printing. Green cloth. VG. **$19.00\£11.00** *(BB 12/88)*

Jackson, J.R. & Bush, F.A. Shrubs and Trees. London, 1947. 1st, 186pp. VG/F in dj. **$7.00\£4.00** *(BLB 7/89)*

Jacob, Dorothy. Flowers in the Garden. NY, 1969. 239pp. VG in dj. **$10.00\£6.00** *(RW 11/86)*

Jacob, Dorothy. Flowers in the Garden—A Personal Reminiscence of Seventy Years of Gardening. NY, 1969. 1st Amer ed. 239pp. VG in slightly trn dj. **$12.00\£7.00** *(BF 2/90)*

Jacob, Rev. Joseph. Daffodils. London & Edinburgh, nd c. 1912. 115pp. Head of spine worn else G. **$18.00\£10.00** *(ABx 3/89)*

Jacobsen, H. Succulent Plants. London, 1955. 4th imp. [xvi], 293pp. Some fdng on brds else VG in torn dj. **$36.00\£20.00** *(Hrt 12/89)*
Another copy. London, 1946. 2nd ed. Ry 8vo. Trifle used. **$27.00\£15.00** *(WW 1/90)*

Jacquin, N.J. Miscellenea Austriaca, Vol. 1. Vienna, 1778. [iv], 212pp. 4to. Calf. Cheap cpy 1 xerox plate, one crudely repaired. Slght margin damage at the beginning; title page is neat MS facsimile on smaller paper. **$216.00\£120.00** *(WW 1/90)*

Jaeger, A. Rosenlexicon. Leipzig, 1960. 768pp. In German. Upper spine defective else G. **$45.00\£25.00** *(PBx 11/89)*

James, Chas. Practical Agriculture. NY, 1908. 203pp. VG. **$8.00\£4.00** *(AB 4/89)*

James, Emma Harvey. Unusual Vegetables. London, 1956. G in dj. **$9.00\£5.00** *(ABx 3/89)*

Janse, J.A. Succulent Plants. Hillegom, 1959. 116pp. G. **$27.00\£15.00** *(WW 1/90)*

Janzen, D.H. Swollen-Dorn Acacias of Central America. 1974. Smithsonian Contr. Bot. 13. 131pp. G. **$14.00\£8.00** *(PBx 11/89)*

(Japanese dendrology) A Quarto volume, in Japanese, with 100 double-page b/w drawings that show the leaf form, blossom, fruit, and wood grain for 100 trees. Text is in Japanese, but each tree is also identified in English.

Bound in is a letter in Japanese, and a translation into English, from Takeii Morimasa, Chief of the Forest Bureau of Japan, dated Nov. 17, 1882 and addressed to F.B. Hough, then chief of the Division of Forestry, U.S. Dept. of Agriculture, thanking him for 2 vols sent to him and enclosing the present vol on the dendrology of Japan. There is a mdrn ownr's inscrip on the frnt e.p. and a paper crack at the rear hinge. Bnd in 3/4 brwn clf, stmpd in gilt, with the title lettering in Japanese. **$300.00\£167.00** *(BF 2/90)*

Jarchow, Merrill. The Earth Brought Forth. St. Paul, 1949. 314pp. VG in lightly worn dj. **$20.00\£11.00** *(AB 4/89)*

Jekyll, Francis. Gertrude Jekyll, a Memoir. London, 1934. 248pp. VG in rbbd clth. **$45.00\£25.00** *(BF 2/90)*

Jekyll, Gertrude & Sir Lawrence Weaver. Gardens for Small Country Houses. London, 1924. 5th ed. 262pp. Sm folio. VG in slightly rbbd and flaked blue clth. **$100.00\£56.00** *(BF 2/90)*

Another copy. London, 1913. 2nd ed. [1], 258pp. Cvrs spttd, 5 pp reprd with tape in mrgns. Fair. **$30.00\£17.00** *(CB 1/90)*

Jekyll, Gertrude & Sir Lawrence Weaver. Gardens for Small Country Houses. London, 1927. 6th ed. 262pp. 4to. G. **$117.00\£65.00** *(ABx 3/89)*

Jekyll, Gertrude. Annuals and Biennials. London, 1916. 174pp. Some fxng, edge-wrn and flaked bndng. Sound. **$55.00\£31.00** *(BF 2/90)*

Jekyll, Gertrude. Colour in the Flower Garden. London, 1908. 148pp. Occas fxng, bkplte else VG in blue clth. **$80.00\£44.00** *(BF 2/90)*

Jekyll, Gertrude. Colour in the Flower Garden. London, 1908. 1st ed. [xiv], 148pp. G. **$86.50\£48.00** *(ABx 3/89)*

Jekyll, Gertrude. Colour Schemes for the Flower Garden. London, 1925. 6th ed. 159pp. Ownr's sig, photo of Jekyll on e.p. Cvrs spttd; one crnr roughened. **$60.00\£33.00** *(BF 2/90)*

Jekyll, Gertrude. Colour Schemes for the Flower Garden. London, 1914. 3rd ed. [xvi], 159pp + 16pp pub. cat. VG. **$65.00\£36.00** *(RW 6/89)*

Jekyll, Gertrude. Colour Schemes in the Flower Garden. London, 1911. 148pp. VG. **$70.00\£39.00** *(BF 2/90)*

Jekyll, Gertrude. Home and Garden. London, 1910. 4th imp. 301pp. E.p.s. brwnd else VG in decor clth. **$65.00\£36.00** *(BF 2/90)*

Jekyll, Gertrude. Home and Garden. London, 1926. Reissue, 301pp. G. **$72.00\£40.00** *(ABx 3/89)*

Another copy. London, 1901. New ed. (3rd rpt). [xv], 301pp. Slght spine fade else VG. **$70.00\£39.00** *(Hrt 12/89)*

Jekyll, Gertrude. On Gardening. NY, 1964. 283pp. F in slightly wrn dj. **$30.00\£17.00** *(BF 2/90)*

Jekyll, Gertrude. On Gardening. NY, 1964. 283pp. VG in dj. **$20.00\£11.00** *(RW 11/86)*

Jekyll, Gertrude. Wall and Water Gardens. London, 1903. 2nd ed. [xiv], 177pp. Orig clth, slightly used, some foxing. **$63.00\£35.00** *(WW 1/90)*

Another copy. London, 1901. 177pp. Mod fxed, bndng rbbd. **$70.00\£39.00** *(BF 2/90)*

Jekyll, Gertrude. Wall and Water Gardens. London, Country Life, nd. 7th ed. [xvi], 214pp. Orig. blue cloth. VG. **$45.00\£25.00** *(RW 11/86)*

Another copy. NY, 1901. 8vo, faded green cloth. 177pp. Some foxing, some pencilled notations, moderate wear. **$45.00\£25.00** *(MD 7/89)*

Jekyll, Gertrude. Wall, Water and Woodland Gardens. London, 1933. 8th ed. 262pp. Clth neatly rebacked. **$72.00\£40.00** *(WW 1/90)*

Jekyll, Gertrude. Wood and Garden. London, 1901. New ed, 9th impr. [xvi], 286pp. Tips bmpd, stmp on pastedwn, overall nice looking, tight. **$45.00\£25.00** *(CB 1/90)*

Jekyll, Gertrude. Wood and Gardens. London, 1899. 6th imp. [xvi], 286pp. Spine fade & tender, 1 plte loose. **$70.00\£39.00** *(Hrt 12/89)*

Jellicoe, G.A. Gardens of Europe. London, 1937. 134pp. F in slightly wrn dj. **$85.00\£47.00** *(BF 2/90)*

Jellicoe, G.A. Gardens of Europe. London, 1937. 1st ed. 134pp. G. **$54.00\£30.00** *(ABx 3/89)*

Jellicoe, Geoffrey & Susan. The Landscape of Man. London, 1975. 383pp. VG in dj. **$70.00\£39.00** *(Hrt 12/89)*

Jellicoe, Geoffrey & Susan. The Landscape of Man. London, 1975. 383pp. VG in dj. **$75.00\£42.00** *(RW 6/89)*

Jellicoe, Susan & Geoffrey. Modern Private Gardens. London, 1968. 127pp. F in dj. **$45.00\£25.00** *(Hrt 12/89)*

Jennings, Arthur & Leonard Johnson. The Complete Home Landscape. NY, 1923. [x], 178pp. Spine fade else VG. **$15.00\£8.00** *(Hrt 12/89)*

Jensen, Jens. Siftings—The Major Portion of the Clearing and Collected Writings. Chicago, 1956. 145pp. VG in dj. **$50.00\£28.00** *(BF 2/90)*

Jepson, W.L. A Flora of California. Berkeley, (1909)-1943. 3 vols. in 4, ry 8vo, wrpprs, unopnd. G. **$99.00\£55.00** *(WW 1/90)*

Jepson, Willis L. A Manual of the Flowering Plants of California. Berkeley & LA. U of Cal Press, 1957. 3rd prntng. 1238pp. Tips wrn, bck cvr scraped else VG. **$35.00\£19.00** *(CB 1/90)*

Jepson, Willis. A Flora of the Economic Plants of California for Agriculture Students. Berkeley, 1924. 223pp. VG. **$10.00\£6.00** *(WH 1/90)*

Jermyn, Stanley T. Flora of Essex. London, 1974. 302pp. G in dj. **$45.00\£25.00** *(ABx 3/89)*

Another copy. London, 1975. Clean, 2nd hnd. **$23.00\£13.00** *(WW 1/90)*

Jex-Blake, Lady M. Some Wild Flowers of Kenya. 1948. In paper as issued. VG. **$23.00\£12.50 (DL 89/90)**

Johns, Rev. C.A. Flowers of the Field. London, 1909. 3rd impr. 324pp. VG. **$45.00\£25.00** *(RW 11/86)*

Johnson, A.T. A Woodland Garden. London, 1937. 232pp. Slght fxng, ownr's bkplte. VG. **$35.00\£19.00** *(BF 2/90)*

Johnson, A.T. Hardy Heaths and Some of Their Nearer Allies. London, 1956. Rev. ed. 127pp. Small 8vo. G. **$7.25\£4.00** *(ABx 3/89)*

Johnson, A.T. The Mill Garden. London, 1950. 207pp. G in torn dj. **$11.00\£6.00** *(ABx 3/89)*

Johnson, A.T. The Mill Garden. London, 1950. Trifle used, spine faded. **$27.00\£15.00** *(WW 1/90)*

Johnson, Clifton. Among English Hedgerows. NY, 1899. 347pp. Decorated cloth, as new. **$15.00\£8.00** *(AB 2/89)*

Johnson, George W. & Robert Hogg. Journal of Horticulture. Vols. 16 & 17. London, 1869. 2 vols. [viii], 538pp, [viii], 436pp. 8xll. G. **$54.00\£30.00** *(ABx 3/89)*

Johnson, Leonard H. Foundation Planting. Delaware, NY, 1927. [xviii], 236pp. VG. **$17.00\£9.00** *(CB 1/90)*

Johnson, Louisa. Every Lady Her Own Flower Gardener. NY, 1855. Rev. frm 14th Lndn ed. [viii], 119pp. Bnd w/American Rose Culturist. NY, 1855. 96pp. Lt foxing else VG in pblshr's clth titled American Florist Guide. **$95.00\£53.00** *(Hrt 12/89)*

Johnson, Loyal B. How to Landscape Your Garden. NY, 1941. 1st ed. [iix], 221pp. VG. **$22.00\£12.00** *(Hrt 12/89)*

Johnson, Rev. W. Nature and Naturalists. London, nd, (1903). [viii], 171pp. Gilt dec brds. G. **$40.00\£22.00** *(Hrt 12/89)*

Johnson, T. Botanical Journeys in Kent and Hampstead. London, 1971. Facs. Reprnt. [x], 167pp. Leathr. G. **$32.00\£18.00** *(WW 1/90)*

Johnson, Thomas, ed by J.S.L. Gilmour. Thomas Johnson Botanical Journeys in Kent & Hampstead. Pitts. The Hunt, 1972. [ix], 167pp. VG. **$40.00\£22.00** *(Hrt 12/89)*

Johnson, Thomas. Iter Plantarum. 1629 (and) Descriptio Itineris Plantarum, 1632. 1972. Hunt Facs. Prod. G. **$18.00\£10.00** *(PBx 11/89)*

Jones & Wolff (ed). Agrarian Change and Economic Development. L., Metheun, 1969. 172pp. Dj worn, else VG. **$10.00\£6.00** *(AB 4/89)*

Jones, George Nevile. A Botanical Survey of the Olympic Peninsula, Washington. Seattle, 1947 (1936). 286pp. Blue cloth, pub. by U. of Wash. Pr. VG. **$38.00\£21.00** *(BB 12/88)*

Jones, Ina. Arranging Church Flowers. Dallas, 1951, 2nd ed. 112pp. As new in dj. **$12.00\£7.00** *(AB 2/89)*

Jones, J.P. & J.F. Kingston. Flora Devoniensis. London, 1829. 2 prts in one vol. New clth. **$90.00\£50.00** *(WW 1/90)*

Jones, K. Botanical Essays for Humanists. Wahr, 1972. 349pp. VG in dj. **$16.00\£8.50 (GW 8/89)**

Jones, Louise S. Put a Feather in Your Hat. LA, 1938. 150pp. Signed. VG in worn dj. **$10.00\£6.00** *(AB 2/89)*

Jones, Louise S. Who Loves a Garden. Primavera Press, 1934. 195pp. Ltd to 500 copies. Light wear. **$12.00\£7.00** *(AB 2/89)*

Jordan, J.M. Jordan's Illustrated Horticultural Guide. St. Louis, 1871. 120pp. Wraps a bit worn, contents as new. **$80.00\£44.00** *(AB 4/89)*

Justice, J. The British Gardener's New Director. Dublin. 1765. 4th ed. [xvi, xxvi], 443, [12]pp. New clth. Upper corner of title page cut away. Slght waterstning at end, tear in one plt reprd. **$108.00\£60.00** *(WW 1/90)*

K

Kains, M.G. Adventures in Gardening. NY, 1938. [viii], 280pp. VG in dj. **$12.50\£7.00** *(RW 6/89)*

Kains, M.G. Culinary Herbs. NY, 1912. 143pp. Moderate wear. **$10.00\£6.00** *(RW 11/86)*

Kains, M.G. Ginseng. NY, 1904. Rev. ed. 144pp. Lightly rubbed else VG. **$25.00\£14.00** *(AB 4/89)*

Kalcas, Evelyn Lyle. Food From the Fields. Izmir, 1974. [x], 146pp. Photo wraps. Signed. VG. **$27.50\£15.00** *(BB 12/88)*

Kanjilal, U. et al. Flora of Assam. Assam, 1940. 377pp. VG. **$12.00\£7.00** *(GW 8/89)*

Karel, Leonard. Dried Flowers. Metuchen, 1973. 1st ed. 184pp. F in dj. **$7.00\£4.00** *(BLB 7/89)*

Karel, Leonard. Dried Grasses, Grains, Gourds, Pods & Cones. Metuchen, 1975. Reprnt. 201pp. As new. **$8.00\£4.00** *(BLB 7/89)*

Karr, Alphonse. A Tour Round My Garden. London, 1865. New ed. 332pp. Decorated cloth. G. **$27.00\£15.00** *(ABx 3/89)*

Kassler, Elizabeth B. Modern Gardens and the Landscape. NY, MOMA, 1964. 96pp. Square 8vo. Green cloth. Nice in slightly worn dj. **$17.50\£10.00** *(MD 7/89)*

Another copy. NY, 1964. 96pp. Wraps. VG. **$15.00\£8.00** *(RW 6/89)*

Another copy. MOMA, 1964. 96pp. Wraps. VG $15.00\£8.00 *(CB 12/89)*

Kauffman, C.H. The Gilled Mushrooms. NY, 1971. Vol. I: [xxxii], 442pp. Vol. II: xi, 443-924pp. Wrps. Ownr's sig else VG. **$36.00\£20.00** *(Hrt 12/89)*

Keays, Mrs. Frederick. Old Roses. NY, 1935. 222pp. Cloth, light fading to margins. Else VG. **$30.00\£17.00** *(AB 4/89)*

Another copy. NY, 1935. 1st. 222pp. Scarce in this edition, minimal cover defects. **$39.00\£22.00** *(BLB 7/89)*

Keble Martin, W. Over the Hills. London, 1968. 173pp. G in dj. **$10.75\£6.00** *(ABx 3/89)*

Keble Martin, W. Sketches for the Flora. London, 1972. 208pp. G. **$21.50\£12.00** *(ABx 3/89)*

Keeler, H. Our Garden Flowers. Scribner, 1910. 550pp. Decorated covers. VG. **$16.00\£9.00** *(GW 12/88)*

Keeler, H.L. Our Northern Shrubs and How to Indentify Them. NY, 1925. Orig clth. Nice cpy. **$27.00\£15.00** *(DL 89/90)*

Keeler, Harriet. Our Native Trees and How to Identify Them. NY, 1900. 2nd ed. [xxiii], 533pp. VG. **$15.00\£8.00** *(RW 6/89)*

Keeler, Harriet. Our Northern Shrubs and How to Identify Them. NY, 1928. [xxxiv], 521pp. Dec cvr. VG in torn dj. **$15.00\£8.00** *(RW 6/89)*

Kelaart, E.F. Contributions to the Botany and Topography of Gibraltar and Its Neighbourhood. 1846. With Plan and Views of the Rock. Orig clth, piece mssng frm tip of spine else well preservd cpy. **$72.00\£40.00** *(DL 89/90)*

Kellerman, W.A. Plant Analysis: A Classified List of the Wild Flowers of the U.S. with Keys. . . Phila, 1884. Decor bndng. VG & crisp. **$18.00\£10.00** *(PBx 11/89)*

Kelley, A.P. Mycotrophy in Plants. Chronica Botanica, 1950. G. **$18.00\£10.00** *(WW 1/90)*

Kelly, Howard. Some American Medical Botanists. Troy, 1914. 215pp. VG in slght tendr gilt stmpd clth. **$85.00\£47.00** *(Hrt 12/89)*

Kelway, Christine. Gardening on Sand. London, 1965. 1st ed. 144pp. G in dj. **$28.75\£16.00** *(ABx 3/89)*

Kemp, Edward. How to Lay Out a Garden. London, 1858. 2nd ed. [xxxii], 403pp. A.e.g. **$55.00\£31.00** *(RW 11/86)*

Kemp, Edward. Landscape Gardening. NY, 1912. 4th ed. [xxii], 292pp. VG. **$30.00\£17.00** *(RW 11/86)*

(Kent, Elizabeth). Flora Domestica. London, 1823. [xxiv], 396pp. Re-backed full calf with original spine laid down. VG. **$176.50\£98.00** *(ABx 3/89)*

Kerner von Marilaun, A. The Natural History of Plants. London, 1895. 2 vols. Half morocco. G. **$81.00\£45.00** *(WW 1/90)*

Kew, Royal Botanic Gardens. Hand-List of Herbaceous Plants. London, 1902. 2nd ed. 1235pp. G. **$32.50\£18.00** *(ABx 3/89)*

Kiaer, E. & V. Hancke. Store Rosenbog. Copenhagen, 1965. 4to. VG in dj. **$30.00\£17.00** *(PBx 11/89)*

Kiaer, Eigil. Concise Handbook of Roses. NY, 1966. 111pp. As New. **$20.00\£11.00** *(AB 4/89)*

Kidd, M.M. Wild Flowers of the Cape Peninsula. London, 1950. G. **$36.00\£20.00** *(WW 1/90)*

Kidder, J. Edward. Japanese Temples. Tokyo, Amsterdam, 1964. Folio, brown cloth. 554pp. F. **$120.00\£67.00** *(MD 11/88)*

Kidder, Jr., J. Edward. Japanese Temples. Tokyo & Amsterdam, 1964. 554pp. Sm folio. As New. **$185.00\£103.00** *(Hrt 12/89)*

Killebrew, J. The Grasses of Tennessee Including Cereals & Forage Plants. American, 1878. 511pp. Worn bindings. **$20.00\£11.00** *(GW 12/88)*

Kimball, Winifred. Riminiscences of Alvan Wentworth Chapman. Reprt of the Jrnl of NY Bot Grdn, 22. 1-12. 1921. 11pp. Wrps. G. **$10.00\£6.00** *(Hrt 12/89)*

Kincaid, Mrs. Paul. The Camellia Treasury. NY, 1964. 224pp. VG in worn dj. **$10.50\£6.00** *(BLB 7/89)*

Kinch, Raymond. Nebraska Weeds. Lincoln, 1939. Rev. ed. Sftbnd, cloth spine. 241pp. Minimal cover defects, light damp mark. **$10.50\£6.00** *(BLB 7/89)*

King, Arthur. The Observer's Book of Garden Flowers. London, 1965. 6th prt. 239pp. Sm 16mmo. VG. **$5.00\£3.00** *(BLB 7/89)*

King, Eleanor. Bible Plants for American Gardens. NY, 1941. 1st, 203pp. VG/F in dj. **$9.25\£5.00** *(BLB 7/89)*

King, G. Materials for a Flora of the Malayan Peninsula. Asiatic Soc., 1905. 386pp. Broken binding. **$22.00\£12.00** *(GW 8/89)*

King, Mrs. Francis. The Flower Garden Day by Day. NY, 1927. 210pp. Cover edge wear else VG. **$5.75\£3.00** *(BLB 7/89)*

King, Mrs. Francis. The Little Garden. Boston, 1922. 94pp. VG. **$7.50\£4.00** *(RW 11/86)*

King, Mrs. Francis. The Well-Considered Garden. NY, 1917. 290pp. F in slightly spttd dec clth. **$18.00\£10.00** *(BF 2/90)*

King, Mrs. Francis. The Well-Considered Garden. NY, 1922. New and rev. ed. [xv], 234pp. VG. **$10.00\£6.00** *(RW 11/86)*

King, Mrs. Francis. Variety in the Little Garden. Boston, 1923. 120pp. VG. **$7.50\£4.00** *(RW 11/86)*

Another copy. Boston, 1923. 120pp. Spine fade else VG. **$16.00\£9.00** *(Hrt 12/89)*

King, T. The Clydesdale Flora. 277pp. Hopkins, 1891. VG. **$20.00\£11.00** *(GW 12/88)*

Kingdon-Ward, Frank. Field Notes of Plants, Shrubs and Trees Other than Rhododendrons Collected in 1924/5. London, 1926. 55pp. Priv prnted. G. **$27.00\£15.00** *(WW 1/90)*

Kingdon-Ward, Frank. Field Notes of Rhododendrons. Privately printed, 1933. 43pp. Scarce. G. **$32.50\£18.00** *(ABx 3/89)*

Kingdon-Ward, Frank. Pilgrimage for Plants. London, 1960. 1st ed. 191pp. G in dj. **$57.50\£32.00** *(ABx 3/89)*

Another copy. London, 1960. 1st ed. 192pp. VG. **$45.00\£25.00** *(JI 12/88)*

Kingdon-Ward, Frank. The Land of the Blue Poppy. Sakonnet, 1973. [xiii], 215pp. As New in dj. **$55.00\£31.00** *(Hrt 12/89)*

Kingdon-Ward, Frank. The Romance of Gardening. London, 1935. 1st ed. 271pp. G. **$43.25\£24.00** *(ABx 3/89)*

Kirby, A.M. Daffodils, Narcissus. NY, 1907. 235pp. VG. **$5.00\£3.00** *(RW 11/86)*

Kirk, Donald. Wild Edible Plants of North America. Happy Camp, CA, 1975. Rev. ed. 342pp. Wraps. 12mo. As New. **$10.00\£6.00** *(JI 4/89)*

Kirk, T. The Forest Flora of New Zealand. Wellington, 1889. [xv], 345pp. Sm folio, orig clth. Bndng trifle used, inscrip on title-pge. **$270.00\£150.00** *(WW 1/90)*

Kirkegaard, John. A Practical Handbook of Trees, Shrubs, Vines and Herbaceous Perennials. Boston, 1916. 407pp. G. **$18.00\£10.00** *(ABx 3/89)*

Kirkwood, J.E. Northern Rocky Mountain Trees and Shrubs. Stanford, 1930. [xvii], 340pp. Green cloth. VG. **$37.50\£21.00** *(BB 12/88)*

Kitagawa, M. Lineamenta Florae Manshuricae. Hsinking, 1939. 488pp. Ry 8vo, orig limp brds (hd of spine trn). Lower mrgn wtr stnd with some cockling of paper. Faint libry stmp on cvr. **$108.00\£60.00** *(WW 1/90)*

Kitamura, F. & Y. Ishizu. Garden Plants. Tokyo, 1963. 266pp. Ownr's stamp else VG. **$45.00\£24.00** *(Hrt 12/89)*

Klaber, Doretta. Rock Garden Plants. NY, 1959. 1st. 185pp. Spine wear else VG. **$9.25\£5.00** *(BLB 7/89)*

Knowlton, Frank Hall. Plants of the Past, A Popular Account of Fossil Plants. Princeton, 1927. [xix], 275pp. Tender brds else VG. **$20.00\£11.00** *(Hrt 12/89)*

Kolaga, Walter. All About Rock Gardens and Plants. NY, 1966. 385pp. VG. **$12.00\£7.00** *(RW 6/89)*
Another copy. NY, 1966. 385pp. VG in dj. **$18.00\£10.00** *(Hrt 12/89)*

Kops, J. Flora Batava. Amsterdam, 1807. 39 hand-colored plates from vol. 2 unbound. **$108.00\£60.00** *(WW 1/90)*

Kostler, J. Silviculture. London, 1956. 432pp. G. **$27.00\£15.00** *(WW 1/90)*

Kraemer, H. Scientific & Applied Pharmacognosy. Author, 1915. 857pp. VG. **$55.00\£31.00** *(GW 12/88)*

Kraft, Ken and Pat. The Best of American Gardening. NY, 1975. 1st ed. 214pp. VG in edge-worn dj. **$8.00\£4.00** *(CB 12/89)*

Kraft, Ken. Garden to Order. NY, 1963. [xii], 324pp. VG in dj. **$5.00\£3.00** *(RW 11/86)*

Kramer, F.L. The White House Gardens. NY, 1973. 1st ed. Oblong large 8vo. 144pp. As New in dj. **$11.75\£7.00** *(BLB 7/89)*
Another copy. NY, 1973. 1st ed. 144pp. G in dj. **$45.00\£25.00** *(ABx 3/89)*
Another copy. NY, 1973. 1st prntng. 144pp. G. **$15.00\£8.00** *(CB 8/89)*

Kramer, Jack. Hanging Gardens. NY, 1971. 101pp. G in dj. **$14.50\£8.00** *(ABx 3/89)*

Kramer, Jack. Natural Gardens, Gardening with Native Plants. NY, 1973. 150pp. VG in dj. **$15.00\£8.00** *(Hrt 12/89)*

Kredel, Fritz (illus). Glass Flowers from the Ware Collection. NY, 1940. 58pp. Brds. **$15.00\£8.00** *(WH 1/90)*

Kreig, Margaret. Green Medicine. Chicago, 1964. 4th prntng. 462pp. VG in worn dj. **$12.00\£7.00** *(CB 12/89)*

Krieger, Louis C.C. The Mushroom Handbook. NY, 1947. 538pp. Front e.p. starting else VG in chipped dj. **$25.00\£14.00** *(JI 12/88)*
Another copy. NY, 1936. 538pp. VG. **$25.00\£14.00** *(WH 1/90)*

Krockmal, Arnold. Guide to Medicinal Plants of Appalachia. GPO, 1969. 291pp. Stiff wraps, As New. **$20.00\£11.00** *(AB 4/89)*

Kruckeberg, Henry W. George Christian Roeding. LA, 1930. 109pp. Fine in dj. **$32.00\£18.00** *(BB 12/88)*

Krutch, Joseph Wood (ed). The Gardener's World. NY, 1959. 2nd imp. 476pp. Green boards. G in dj. **$39.50\£22.00** *(ABx 3/89)*
Another copy. NY, 1959. 476pp. Green cloth. Ownr's inscrp. f.f.e.p. else VG in tattered dj. **$20.00\£11.00** *(CB 8/88)*

Kuck, L.F. & R.C. Tongg. Hawaiian Flowers. Honolulu, 1943. 1st, lg 8vo. VG. **$8.00\£4.00** *(BLB 7/89)*

Kuck, Loraine E. & Richard C. Tongg. The Modern Tropical Garden. Honolulu, 1958. 4to. 250pp. Tan cloth. VG in chipped dj. **$25.00\£14.00** *(BB 12/88)*

Kuhner, F. & H. Romagnesi. Flore Analytique des Champignons Superieurs. Paris, 1974. [xiv], 557pp. G. **$111.00\£62.00** *(WW 1/90)*

Kuhner, R. & H. Romagnesi. Flore Analytique des Champignons Superieurs (Agarics, Boleta, Chanterelles). 1053. Orig ed. VG. **$81.00\£45.00** *(DL 89/90)*

Kullenberg, B. Studies in Orphys Pollination. Uppsala, 1961. 340pp. Wrps. G. **$72.00\£40.00** *(WW 1/90)*

Kumlein, L.L. The Friendly Evergreens. Dundee, 1946. 1st, sm 4to. 237pp. VG. **$17.75\£10.00** *(BLB 7/89)*
Another copy. NY, 1954. 237pp. F in worn dj. **$15.00\£8.00** *(WH 1/90)*

Kummerly, Walter (ed). The Forest. Wash, 1973. 1st, sm folio. 299pp. F in G dj. **$15.25\£8.00** *(BLB 7/89)*

Kunth, C.S. Enumeratio Plantarum Omnium Hucusque Cognitarum. Vol. 3. Stuttgart, 1841. 644pp. 8vo. Wrappers (neatly rebacked), uncut. Old lib. stmp title page. **$63.00\£35.00** *(WW 1/90)*

Kunzle, Father Joh. Herbs and Weeds. Locarno, 1967. 79pp. Light pencil notes else VG in wraps. **$24.00\£13.00** *(Hrt 12/89)*

Kupper, W. Kakteen. Zurich, 1954. 128pp. G. **$36.00\£20.00** *(WW 1/90)*

Kupper, W. Orchids. London, 1961. 128pp. Ry 8vo. G. **$30.00\£17.00** *(WW 1/90)*

Kurylowicz, W. Atlas of Spores of Selected Genera & Species of Streptomycetaceae. Univ. Park, 1971. 110pp. Folio. VG in dj. **$12.50\£7.00** *(GW 8/89)*

Kyle, Frank. Chrysanthemums. London, 1961. 4th prntng. F in dj. **$7.75\£4.00** *(BLB 7/89)*

L

L'Heritier de Brutelle, Charles-Louis. Sertum Anglicum, 1788. Pitts, 1963. Facs ed. Edition of 2000. F. **$30.00\£17.00** *(BF 2/90)*
Another copy. Pitts, 1963. Ltd. to 2000. VG. **$30.00\£17.00** *(CB 1/90)*

L'Hertier, Charles-Louis. Sertum Anglicum. Hunt Bot. Lib., 1788, 1963. 1st thus. Folio, 140pp. 2000 copies, As New. **$20.00\£11.00** *(BLB 7/89)*
Another copy. 1788 (1963). Ltd. ed of 2000. F. **$22.00\£12.00** *(PBx 11/89)*

(La tourette, M.A.L.C. de) Demonstrations Elementaires de Botanique. Lyon. 1787. 3 vols. complete. fll clf. Ownr's sig. (Luzenberg). Extremities wrn, scffd., but tight, sound. Contents brwning. **$250.00\£139.00** *(PBx 11/89)*

La Quintinye (J. De.). Le Parfait Jardinier, ou, Instruction pour les Jardins Fruitiers et Potagers avec un Traité des Orangers. . .Paris. 1695. Sm folio, fll clf. 11 plts & 9 vignettes. 2nd ed., compl., 2 vols in 1 (bnd with) Instruction pour la Culture des Fleurs. Paris. 1695. 846pp. Extremities rbbd; spine chppd else solid, tight. F. **$1,075.00\£597.00** *(PBx 11/89)*

Labyrinte de Versailles. Suivant la Copie, a Paris, de l'Imprimerie Royale (1963). A Amsterdam, Chez Pierre Mortier avec Privelege. Title w/royal coat of arms, blank, (3) of descrip, (1) explanation of labyrinth plan, (1) plan, blank, (1) entrance, followed by the 39 plts w/facing descrip & verse, (1) table of fables. Wrn hlf clf, frnt brd loose, portion of spine's small title label, glt on blck, missing. Rare. **$2150.00\£1194.00** *(Hrt 12/89)*

Laing, R.M. & E.W. Blackwell. Plants of New Zealand. Christchurch, 1964 (1906). 7th ed. rev. [xvii]. VG in dj. **$46.00\£26.00** *(BB 12/88)*

Lamarck, J.B. de and A.P. Candolle. Flore Francaise. Paris, 1815. 3rd ed. 6 vols. Lib. bckrm (with stmps). **$396.00\£220.00** *(WW 1/90)*

Lamb, E. & B. Pocket Encyclopedia of Cacti in Colour. London, 1969. G. **$22.00** *(WW 1/90)*

Another copy. London, 1969. Sm 8vo. 217pp. VG in bright, tattered dj. Signed by the authors. **$15.00\£27.00** *(CB 1/90)*

Lamb, Edgar & Brian. Colorful Cacti of the American Deserts. NY, 1974. 1st US. Sm 12mo. 236pp. F in dj. **$7.00\£4.00** *(BLB 7/89)*

Lamb, Edgar & Brian. Pocket Encyclopedia of Cacti in Colour. London, 1971. 4th prntng. 217pp. VG in torn dj. **$8.00\£4.00** *(BLB 7/89)*

Another copy. London, 1969. 217pp. Signed by authors. VG in torn dj. **$13.00\£7.00** *(CB 8/89)*

Lambert, Audrey M. The Making of the Dutch Landscape. London, 1971. 412pp. Bookplt, else nr F. **$30.00\£17.00** *(Hrt 12/89)*

Lamborn, L.L. American Carnation Culture. Alliance, 1892. 3rd ed. 217pp. Library blind stamps, scarce. Nr F. **$11.75\£7.00** *(BLB 7/89)*

Lamplugh, Anne. My Flower Decoration Notebook. London, 1937. 1st. 103pp. VG. **$8.00\£4.00** *(BLB 7/89)*

Lamson, Mary. Gardening with Shrubs and Small Flowering Trees. NY, 1946. 1st. 295pp. Minor cover defects else VG. **$7.00\£4.00** *(BLB 7/89)*

Lancaster, Osbert. Classical Landscape with Figures. Boston, 1948. 224pp. VG in older dj. **$42.00\£23.00** *(Hrt 12/89)*

Landing, James. American Essence. Kalamazoo, 1969. 244pp. As New in dj. **$45.00\£25.00** *(AB 4/89)*

Landsell, J. Grapes, Peaches, Nectarines and Melons. London, nd (1939?). 144pp. G. **$18.00\£10.00** *(ABx 3/89)*

Lane, Ferdinand. All About the Flowering World. NY, 1956. 3rd prntng. 141pp. Damp mks bottom of some pp else VG in dj. **$5.75\£3.00** *(BLB 7/89)*

Langdon, Rev. S. My Mission Garden. London, 1891. 2nd ed. Sm 8vo. 200pp. Sl tender in gilt/blck dec brds/spine, some fxing else G. **$30.00\£17.00** *(Hrt 12/89)*

Lange, J. Ordbog Over Danmarks Plantenavne. 3 vols. Copenhagen, 1959/61. 4to. VG. **$120.00\£67.00** *(PBx 11/89)*

Lanjouw, J. & F.A. Stafleu. The Herbaria of the World. 5th ed. 1964. 251pp. G. **$18.00\£10.00** *(PBx 11/89)*

Lankester, Mrs. Talks About Plants, or Early Lessons in Botany. London,

1878. 252pp. As New. **$25.00\£14.00** *(AB 2/89)*

Lape, Fred. A Garden of Trees and Shrubs. Ithaca, 1965. [xii], 129pp. VG in dj. **$45.00\£25.00** *(Hrt 12/89)*

Large, E.C. The Advance of the Fungi. London, 1940. 488pp. G. **$45.00\£25.00** *(WW 1/90)*

Laurence, Jeanne. Album of Alaska Wild Flowers. Typed, bound manuscript. (Book pblshd 1974). Nd. 93 fll pg orig photos of wild flower paintings. Text pp gilt-ruled by hand. G. **$75.00\£42.00** *(PBx 11/89)*

Laurie, Alex & D. Kiplinger. Commercial Flower Forcing. Philadelphia, 1958. 6th ed. 509pp. Near F. **$8.00\£4.00** *(BLB 7/89)*

Lawrence, Elizabeth. Gardens in Winter. NY, 1961. 1st ed. 218pp. Some pencil notes else VG in dj. **$7.50\£4.00** *(RW 11/86)*

Another copy. 240pp. Claitor, 1961. VG in dj. **$12.50\£7.00** *(GW 12/88)*

Lawrence, Elizabeth. The Little Bulbs. NY, 1957. 1st ed. 248pp. G in soiled, chppd dj. **$15.00\£8.00** *(CB 1/90)*

Lawrence, G. An Introduction to Plant Taxonomy. Macmillan, 1955. 179pp. VG. **$12.50\£7.00** *(GW 12/88)*

Lawrence, G.H.M. (ed). The Bicentennial of Michael Adanson's "Famille des Plantes". Pitts, 1963-4. 2 vols. Vol I: [xii], 392pp. Vol. II: [xii], 243pp. As New in gilt blue clth. **$36.00\£20.00** *(Hrt 12/89)*

Lawrence, George H. Catalogue 2nd International Exhibition of Botanical Art and Illustration. Pittsburgh, 1968. 267pp. VG. **$20.00\£11.00** *(RW 11/86)*

Lawrence, George H.M. (ed). Huntia 1. Pittsburgh, 1964. [vi], 222pp. Wraps. As New. **$40.00\£22.00** *(JI 7/88)*

Lawrence, George H.M. B-P-H. Botanico-Periodicum-Huntianum. Pttsbrgh, 1968. 1063pp. Green clth, As New. **$25.00\£14.00** *(Hrt 12/89)*

Lawrence, George H.M. Taxonomy of Vascular Plants. NY, 1951. 823pp. G. **$22.00\£12.00** *(BB 12/88)*

Another copy. NY, 1958. 21st prntng. 823pp. F. **$10.00\£6.00** *(BLB 7/89)*

Lawrence, George. The Rachel McMasters Miller Hunt Botanical Library. Pittsburgh, 1961. 1st ed. [viii], 44pp. 8vo. Wraps, VG. **$28.00\£16.00** *(JI 4/89)*

Lawrence, W.J.C. & J. Newell. Seed and Potting Compost. London, 1939. 1st, 128pp. VG in dj. **$6.50\£4.00** *(BLB 7/89)*

Another copy. London, 1950. 4th ed., rev. 166pp. VG in worn dj. **$12.00\£7.00** *(AB 2/89)*

Lawrence, W.J.C. Practical Plant Breeding. London, 1942. 155pp. VG/F in dj. **$7.25\£4.00** *(BLB 7/89)*

Lawson, Alexander H. Bamboos. NY, 1968. 192pp. VG in dj. **$65.00\£36.00** *(BB 12/88)*

Lawson, Peter and Son. The Lawsonian Collection. Edinburgh, 1852. [xx], 138, 128, 64, 80, 72, 83pp. Engraved frontis new end papers. Private press of Peter Lawson. VG. **$135.00\£75.00** *(ABx 3/89)*

Lawson, William & C.D. Hunter, et al. Ten Years of Gentleman Farming at Blennerhasset. London, 1874. [viii], 408pp. Some shelfwr else VG. **$30.00\£17.00** *(Hrt 12/89)*

Lawson, William. A New Orchard and Garden. 1927. Ltd ed. of 650 cpies. Cpy #581. VG. **$54.00\£30.00** *(DL 89/90)*

Lawson, William. A New Orchard and Garden. London. Cresset Press, 1927. Reprint. #278 of 650 copies. 116pp. VG in dj. **$45.00\£25.00** *(RW 11/86)*

Lay, Charles Downing. A Garden Book for Autumn and Winter. NY, 1924. 303pp. Glt stampd brds. VG. **$32.00\£18.00** *(Hrt 12/89)*

Lay, Charles Downing. A Garden Book for Autumn and Winter. NY, 1924. 303pp. VG. **$20.00\£11.00** *(RW 11/86)*
Another copy. NY, 1924. Slit bottom spine else VG. **$30.00\£17.00** *(Hrt 12/89)*

Le Blond, Mrs. Aubrey. The Old Gardens of Italy. London/NY, 1912. 173pp. Pict clth, somewhat faded else reasonably G. **$30.00\£17.00** *(Hrt 12/89)*
Another copy. NY, 1926. 1st Amer ed. 172pp. VG in slightly rbbd clth. **$25.00\£14.00** *(BF 2/90)*

Le Blond, Mrs. Aubrey. The Old Gardens of Italy. London, 1926. 2nd ed. rev. 173pp. VG. **$30.00\£17.00** *(RW 6/89)*

Le Gallienne, Richard. The Romance of Perfume. NY, 1928. Missing Hudnut brochure. Front brd loose else VG. **$45.00\£25.00** *(Hrt 12/89)*

Le Maout, E. & J. Decaisne. A General System of Botany. Transl. from the original by Mrs. Hooker... arranged. . . by J.D. Hooker, director of the Botanical Gardens. Kew, 1876. 1,067pp. Very lght moisture rippling; in later, immaculate bndng. **$65.00 £36.00 PBx 11/89)**

Le Petit, A. Fleurs, Fruits and Legumes du Jour. Paris, March 1871. 4to. Orig. brds with col. title lble. Edge reinfor., endpaper replacement. VG. **$175.00\£97.00** *(PBx 11/89)*

Le Rouge, George Luis. Jardins Anglo Chinois, Paris, Chez le Sieur le Rouge, 1776-1788. 19 of 21 Cahiers, oblg 4to, (460mmx305mm) except for 5, (280mmx213mm). Cahiers I, 1-23 plts, 24-26 text, 2 plts. unnumbered, (slght discoloration plt 1). II, 1-23 plts. III, 1-28 plts. IV, 1-30 plts (slght discoloration plts 1-3) & (Pencilled heads around on plt 11). V, (the William Chambers book), title & text, 30pp, 1-20 dbj pp plts, (Lght fxing some plts. VI, 1-30 plts. VII, 1-19 plts, 20-23, text, 24-27 plts. VIII, 1-28 plts. (1 fldng plt counted as 10-11), (dmp stn bottom right crnr plts 3-6. IX, 1-15 plts. X, 1-11 plts, (slght tr & discoloration on fldg plt 2). XI, 1-20 plts. XII, 2-24 plts., (fxng on mended fldg plt. 11). XIII, 1-26 plts. XIV, 1-11 plts., (discoloration & slght tr on fld of plt 1). XV, 1-28 plts. XVI, 1-30; XVII, 1-30plts. XVIII & XIX, 1 plt, 2-3 text, 4-49 plts, (soil on plt 49). Cahiers 4,6,7,8,10 & 12 are bnd in two clth bndngs later period, balance in orig blue marbled wrpprs, except vol V, bnd in similar blue marbld paper on brds. Except for noted faults exceedingly clean copy. Rare. **$14,000.00\£7778.00** *(Hrt 12/89)*

Leavitt, Robert Greenleaf. The Forest Trees of New England. Jamaica Plains, 1933. [viii], 179pp. Sm chppng to spine else about F. **$33.00\£18.00** *(Hrt 12/89)*

Lee, Frederick. The Azalea Book. Van Nostrand, 1958. 324pp. G in dj. **$34.00\£19.00** *(ABx 3/89)*

Leeming, John. The Book of the Delphinium. London, 1932. [ix], 76pp. VG. **$7.50\£4.00** *(RW 11/86)*

Lees, Carlton B. Gardens, Plants and Man. Englewood Cliffs, NJ, 1970. 251pp. VG in trn dj. **$14.00\£8.00** *(BF 2/90)*

Lees, Carlton. Gardens, Plants and Man. NY, 1970. Sm 4to, 251pp. F in dj. **$8.00\£4.00** *(BLB 7/89)*
Another copy. Rutledge, 1970. 251pp. F in dj. **$20.00\£11.00** *(WH 1/90)*

Lees, F.A. The Flora of West Yorkshire. London, 1888. [xii], 843pp. Buckram, G. **$81.00\£45.00** *(WW 1/90)*

Leese, Sir Oliver. Cacti and Other Succulents. London, 1966. 2nd. Narrow 8vo. 112pp. VG in dj. **$7.75\£4.00** *(BLB 7/89)*

Leese, Sir Oliver. Cacti. London, 1973. 4to, 144pp. VG in dj. **$8.00\£4.00** *(BLB 7/89)*

Leighton, Ann. Early American Gardens "For Meate or Medicine". Boston, 1970. 441pp. F in dj. **$42.00\£23.00** *(BF 2/90)*

Leighton, Ann. Early English Gardens in New England. (Boston, 1970.) London, 1970. 464pp. G. **$27.00\£15.00** *(WW 1/90)*

Another copy. London, 1970. 1st English ed. of "Early American Gardens". 441pp. F in slightly wrn dj. **$40.00 £22.00 BF 2/90)**

Leighton, Ann. Early English Gardens in New England. London, 1970. 441pp. G in dj. **$45.00\£25.00** *(ABx 3/89)*

Leighton, Clare. Four Hedges, A Gardener's Chronicle. NY, 1935. 167pp. Dj worn, spine worn. Contents VG. **$18.00\£10.00** *(AB 4/89)*

Another copy. London, 1935. 1st ed. 167pp. G in dj. **$45.00\£25.00** *(ABx 3/89)*

Leighton, M. et al. Vascular Plants of Illinois. Illinois, 1955. 593pp. VG. **$20.00\£11.00** *(GW 8/89)*

Leighton, W.A. A Flora of Shropshire. Shrewsbury, 1841. [xii], 573pp. Hlf morocco, gilt. G. **$90.00\£50.00** *(WW 1/90)*

Lemmon, C. & R. Sherman. Flowers of the World. Doubleday, 1964. 280pp. VG in dj. **$20.00\£11.00** *(GW 12/88)*

Lemmon, Kenneth. The Covered Garden. London, 1962. 284pp. VG. **$20.00\£11.00** *(RW 11/86)*

Another copy. London, 1962. 284pp. 1st ed. VG in tattered dj. **$21.00\£12.00** *(CB 8/89)*

Lemmon, R.S. & C.C. Johnson. Wildflowers of North America. NY, 1961. 280pp. Becoming scarce. VG. **$14.50\£8.00** *(BLB 7/89)*

Leonard, E. Acanthaceae of Colombia. Pt III only. USNH, 1958. 458pp. Wrs. VG. **$10.00\£6.00** *(GW 12/88)*

Another copy. Vols I, II, III. Wash., 1951, 1953. 1st edns. 781pp. Orig wrappers. F. **$65.00\£36.00** *(WH 1/90)*

Leroy, Andre. Dictionnaire de Pomologie. 1867. 6 vols. in 4. 1/2 green clf, a little rbbd but a VG set. Very scarce. **$540.00\£300.00** *(DL 89/90)*

Leschallas, Capt H.P. The Small Alpine Garden; The Small Alpine Garden, Part II. London, nd (1937 & 1938). 2 vols. 79, 33pp. Ills paper covers. G. **$27.00\£15.00** *(ABx 3/89)*

Lester, Francis. My Friend the Rose. Harrisburg, 1942. 90pp. Green cloth, very light wear. **$16.00\£9.00** *(AB 4/89)*

Levison, J.J. The Home Book of Trees and Shrubs. NY, 1940. [xiv], 424pp. VG. **$25.00\£14.00** *(RW 6/89)*

Levy, Juliette de Bairacli. Herbal Handbook for Everyone. Newton, 1967. 178pp. VG in dj. **$20.00\£11.00** *(Hrt 12/89)*

Lewis, Dr. G. Griffen. The Book of Roses. Boston, 1931. Signs. of J.H. Nicolas and E.S. Boerner. VG. **$35.00\£19.00** *(PBx 11/89)*

Lewis, Frank. More Flower Arrangements. London, 1950. 27 b&w studies. E.p.s. fxd, else VG in trn dj. **$22.00\£12.00** *(CB 1/90)*

Lewis, L. The Vegetation of the Shackleford Bank (NC). NC, 1917. 40pp. Wrs. VG. **$12.00\£7.00** *(GW 12/88)*

Leyel, Mrs. C.F. (arranged by). Culpeper's English Physician and Complete Herbal. London, 1947. 128pp. VG in dj. **$25.00\£14.00** *(Hrt 12/89)*

Leyel, Mrs. C.F. Compassionate Herbs. London, 1946. 1st ed. 224pp. 8vo. G. **$29.00\£16.00** *(ABx 3/89)*
Another copy. London, 1946. 2nd imp. 224pp. VG in older dj. **$30.00\£17.00** *(Hrt 12/89)*

Leyel, Mrs. C.F. Hearts-Ease. London, 1949. 333pp. G in dj. **$36.00\£20.00** *(ABx 3/89)*

Li, H. Trees of Pennsylvania, the Atlantic States & the Lake States. PA, 1972. 276pp. Wrs. VG. **$7.50\£4.00** *(GW 12/88)*

Li, H.L. The Garden Flowers of China. NY, 1959. [vi], 240pp. Near F in slightly worn dj. **$53.00\£29.00** *(BB 12/88)*

Lid, J. Norsk Flora. Oslo, 1952. 2nd ed. 771pp. G. **$45.00\£25.00** *(WW 1/90)*

Liesveld, J.H. The Retail Florist. NY, 1951. 1st, 393pp. VG in worn dj. **$5.50\£3.00** *(BLB 7/89)*

Liljeblad, S. Utkast Till en Svensk Flora. Uppsala, 1816. [x], [lviii], 761pp, [2]pp. New cloth, sm stain on first 5 leaves. **$90.00\£50.00** *(WW 1/90)*

Lincoln, Mrs. Almira H. Familiar Lectures on Botany. NY, 1845. New ed., rev. & enlrgd. 220pp. Frnt contemp brd quite tndr, scttrd fxng else reasonably G. **$27.00\£15.00** *(Hrt 12/89)*

Lind, E.M. & A.C. Tallantire. Some Common Flowering Plants of Uganda. London, 1962. 264pp. G. **$36.00\£20.00** *(WW 1/90)*

Lindley, J. The Theory and Practice of Horticulture. London, 1855. 2nd ed. of Theory of Horticulture. [xvi], 606pp. Orig clth, slightly faded, sm tear in joint. **$63.00\£35.00** *(WW 1/90)*

Linnaeus, C. & Frederick Hazsselquist. Iter Palestinum Eller Resa til Heliga Landet. Stockholm. 1757. Contemp 1/2 sheep, extrems wrn. Else a G cpy. **$270.00\£150.00** *(DL 89/90)*

Linnaeus, C. Flora Dalekarlica. Orebro, 1873. [viii], 130pp. Facs; orig wrpprs. Privately printed. G. **$63.00\£35.00** *(WW 1/90)*

Linnaeus, C. Genera Plantarum Eorumque Characters Naturales Secundum Numerum, Figurum, Situm & Proportionem. Leyden. 1742. 2nd ed. Contemp 1/2 sheep, spine wrn & reprd at top/base. **$450.00\£250.00** *(DL 89/90)*

Linnaeus, C. Hortus Cliffortianus. London, 1968. 532pp. 4to. Clth. Facsimile reprnt. Slghtly reduced from orig format. G. **$90.00\£50.00** *(WW 1/90)*

Linnaeus, C. Hortus Upsaliensis, Exhibens Plantas Exoticas, Horto Upsaliensis Academiae. Amsterdam, 1748. 1 sm mrgnal tear, some sptting of margns. Later 1/2 morocco, spine rbbd and faded. **$360.00\£200.00** *(DL 89/90)*

Linnaeus, C. Mantissa Plantarum. London, 1964. [viii], 588pp. 2 prts in 1 vol. G. **$45.00\£25.00** *(WW 1/90)*

Linnaeus, C. Oratio de Necessitate Peregrinationum intra Patriam... Accedunt Johannis Browallii. . . Plantarum Sexuale, et Johannis Gesneri... Disserationes de Partiu Vegetationis et Fructificationes Structure. Leyden 1743. 4 works as called for. Mdrn 1/2 vellum else a very well preservd cpy. **$360.00\£200.00** *(DL 89/90)*

Linnaeus, C. Philosophia Botanica. Vienna, 1755. 364pp. Re-issue of the 1st ed. of 1751. New brds. Trifle foxed. **$108.00\£60.00** *(WW 1/90)*

Linnaeus, C. Systema Vegetabilium. Goettingen, 1825-27. 4 vols (of 5). Vols 1-3 clf-backed brds (rbbed). Vol 4 new clth to match. Last 13 leaves of vol 4 are xeroxed. **$216.00\£120.00** *(WW 1/90)*

Lister, A. A Monograph of the Mycetozoa. British Museum, 1911. [v], 302pp. Sm lib stmp on e.p. & title pg. **$108.00\£60.00** *(WW 1/90)*

Little, E.L. Check-List of Native and Naturalized Trees of the U.S. Including Alaska. (USDA hndbk #41). 1953. 470pp. F. **$16.00\£9.00** *(PBx 11/89)*

Little, Elbert & Frank Wadsworth. Arboles Communes de Puerto Rico y Las Islas Virgenes. Puerto Rico, 1967. 1st, 827pp. F in dj. **$46.00\£26.00** *(BLB 7/89)*

Little, Elbert L. Check List of Native and Naturalized Trees of the United States. Wash., 1953. 472pp. VG. **$15.00\£8.00** *(WH 1/90)*

Lloyd, Christopher. The Well-Tempered Garden. NY, 1971. 1st US, 478pp. VG in dj. **$30.00\£17.00** *(Hrt 12/89)*

Lock, G. Tropical Agriculture Series: Sisal. Longman, 1969. 365pp. VG in dj. **$16.00\£9.00** *(GW 12/88)*

Lockwood, Alice G.B. Gardens of Colony and State. NY. 2 vols, 1931 & 1934. Vol I: [xxii], 464pp. Vol II: [xxi] (i), 444pp. Folio. F. **$600.00\£333.00** *(Hrt 12/89)*

Logan, Mary. The Loganberry. Oakland. Priv. print, 1955. 20pp. Scarce. F. **$7.75\£4.00** *(BLB 7/89)*

Logsdon, Gene. Gardener's Guide to Better Soil. Emmaus, 1975. 246pp. VG in dj. **$8.00\£4.00** *(AB 2/89)*

Lohndorff, Ernst. The Forest of Fear. London, 1956. 1st Englsih, 182pp. Book cover marks else G. **$11.75\£7.00** *(BLB 7/89)*

Lomas, Charlotte R. Garden Whimseys. NY, 1923. 1st ed. [vii], 171pp. Inscribed by author. VG. **$12.50\£7.00** *(RW 6/89)*

Lommasson, Robert. Nebraska Wild Flowers. Lincoln, 1973. 1st ed. [x], 190pp. VG in sl worn dj. **$25.00\£14.00** *(JI 4/89)*

Longyear, William. How to Make Garden Pools. NY, 1936. 96pp. VG in dj. **$20.00\£11.00** *(AB 4/89)*

Lord, Eleanor L. A Gardener's Pick-Up Book. Springfield, MA, 1943. Privately printed. [x], 91pp. VG. **$15.00\£8.00** *(CB 12/89)*

Lorimer, Mary. Among the Trees. NY, 1869. 1st, 153pp. Near F. **$17.75\£10.00** *(BLB 7/89)*

Lotsy, J.P. Vortraege ueber Botanische Stammesgeschichte. 3 vols. 1907-1911. 2,782pp. W.B. Turill's cpy. G. **$85.00\£47.00** *(PBx 11/89)*

Loudon, J.C. (ed). The Landscape Gardening and Landscape Architecture of the Late Humphry Repton, Esq. A new ed. London, 1840. 2pp. advts, [xxxi], 619pp. 20pp. advts. F in bumpd orig clth, spine fade. **$625.00\£347.00** *(Hrt 12/89)*

Loudon, J.C. An Encyclopedia of Gardening. London, 1822. 1st ed. [xviii], 1469pp. First few pp with edges browned, occasional light foxing half brown morocco. Good sound copy. **$176.00\£98.00** *(ABx 3/89)*

Loudon, J.C. An Encyclopedia of Plants. London, 1829. [xx], 2, 1159pp. Few pp at end taped in. Rebckd clf. G. **$99.00\£55.00** *(WW 1/90)*

Loudon, J.C. An Encyclopedia of Trees and Shrubs. London, 1842. 1162pp. Rebound in 1/4 green cloth and grey boards. G in modern binding. **$162.00\£90.00** *(ABx 3/89)*

Loudon, J.C. Arboretum et Fruticetum Britannicum. London, 1844. 2nd ed. 8 vols. Orig clth. Usual fair amount of foxing. G. **$270.00\£150.00** *(WW 1/90)*

Loudon, Mrs. The Amateur Gardener. Rev. by Robinson. Nd (1875?). 192pp. Pop. Warne ed., some browning & scuffing. **$16.00\£9.00** *(PBx 11/89)*

Louis-Marie, P. Flore-Manual de la Province de Quebec, Canada. Montreal, 1931. 321pp. Ry 8vo. G. **$54.00\£30.00** *(WW 1/90)*

Lounsberry, A. A Guide to the Wild Flowers. Stokes, 1899. 347pp. VG. **$20.00\£11.00** *(GW 12/88)*

Lounsberry, Alice. A Guide to the Trees. NY, 1900. 2nd ed. [xvii], 313pp. Head/foot of spine slightly wrn else VG. **$23.00\£13.00** *(CB 1/90)*

Lounsberry, Alice. Gardens Near the Sea. NY, 1910. [xv], 274pp. VG. **$40.00\£22.00** *(RW 11/86)*

Lounsberry, Alice. The Garden Book for Young People. NY, 1908. [xi], 290pp. VG. **$15.00\£8.00** *(RW 11/86)*

Lousley, J.E. (ed). The Changing Flora of Britain, 1953. Species Studies in the British Flora, 1955. Progress in the Study of the British Flora, 1957. 3 vols. Clth. G. **$45.00\£25.00** *(WW 1/90)*

Lousley, J.E. Flora of the Isles of Sicily. London, 1971. G. **$27.00\£15.00** *(WW 1/90)*

Lowe, E.J. Beautiful Leaved Plants. 1861. 1st ed. Ry 8vo. Gilt decor bdg. Well preserved. **$175.00\£97.00** *(PBx 11/89)*

Lowe, E.J. Ferns: British and Exotic. 2nd & last ed. 1872. Vols 1,2,4,5,6,7,8. Orig gilt dec. bdg. Few spine ends chipped else hardly worn. **$275.00\£153.00** *(PBx 11/89)*

Lowe, R.T. A Manual Flora of Madeira and the Adjacent Islands of Porto Santo and the Desertas. London, 1857-1868-1872. Vols I & II, prt 1. [xii], 618, 113pp in 1 vol. New clth. Very scarce. **$135.00\£75.00** *(WW 1/90)*

Lowell, Guy. American Gardens. Boston, 1902. Unpaginated. Folio. Edges lghtly wtrstnd, not affecting plts. Some pp roughly opened. Bound in pict clth; wrn & soiled. Bit shaken but sound. **$125.00\£69.00** *(BF 2/90)*

Lowell, Guy. Smaller Italian Villas & Farmhouses. NY, 1916. 125pp. Some finger smudges else VG in slightly shlfwrn, gilt stmpd clth. **$420.00\£233.00** *(Hrt 12/89)*

Lubbock, J. On British Wild Flowers. London, 1897. [xvi], 194pp. G. **$27.00\£15.00** *(WW 1/90)*

Lubbock, Sir J. A Contribution to Our Knowledge of Seedlings. Orig. ed. 1892. 2 vols. Well preservd. **$68.00\£37.50 (DL 89/90)**

Lucas, S. The Marine Algae of E. Greenland. Part I. Taxonomical Part. Kobenhaven, 1959. 247pp. Wrs. VG. **$15.00\£8.00** *(GW 8/89)*

Lund, S. The Marine Algae of East Greenland. Prts I & II. Kobenhaven, 1959. 317pp. Wrs, VG. **$25.00\£14.00** *(GW 12/88)*

Lutz, L. Traite de Cryptogamie. Paris, 1942. [vi], 586pp. Wrpprs, trifle used. **$21.00\£12.00** *(WW 1/90)*

Lyell, K.M. A Geographical Handbook of All the Known Ferns. London, 1870. [x], [ii], 225pp. Cr 8vo. Orig clth, trifle used. A little fxng beginning/end. **$36.00\£20.00** *(WW 1/90)*

Lynch, Kenneth. Garden Ornaments. Connecticut, nd (1974?). 768pp. 9x14. G. **$36.00\£20.00** *(ABx 3/89)*

Lysaght, A.M. Joseph Banks in Newfoundland and Labrador. 1766, Berkeley, 1971. Folio, 512pp. As New in dj. **$35.75\£20.00** *(BLB 7/89)*
Another copy. Berkeley, nd. 1st US ed. 512pp. As New. **$35.00\£19.00** *(JI 12/88)*

Lysaught, A.M. Joseph Banks in Newfoundland & Labrador 1766. 1971. Unused in dj. **$65.00\£36.00** *(PBx 11/89)*

M

Macauley, Rose. Pleasure of Ruins. London, 1966. 2nd imp. 286pp. Folio. VG in dj. **$85.00\£47.00** *(Hrt 12/89)*

MacDonald, Hector. Native Manitoba Plants. Np, 1968. Softbound, 100pp. Minor cover defects. G/VG. **$5.75\£3.00** *(BLB 7/89)*

MacDonald, Norman. The Orchid Hunters. London, 1940. 282pp. G. **$45.00\£25.00** *(ABx 3/89)*
Another copy. London, nd (1930s). 286pp. A little foxing else near F. **$45.00\£25.00** *(JI 12/88)*

Macfie, D.T. Lilies for the Garden and Greenhouse. London, 1950. New ed. (1939). 151pp. G in dj. **$14.50\£8.00** *(ABx 3/89)*

MacGinitie, H. Early Middle Eocene Flora. . . NW Wind River Basin, Wyoming. CA, 1974. 103pp. Wrs. VG. **$11.00\£6.00** *(GW 12/88)*

MacGregor, Jessie. Gardens of Celebrities and Celebrated Gardens in and around London. London, 1918. 326pp. Remains of bkplte, paper break to rear hnge, else VG in glt-stmpd green clth. Mildly stnd, some damage to spine ends, borders. **$38.00\£21.00** *(BF 2/90)*

MacGregor, Jessie. Gardens of Celebrities and Celebrated Gardens in and around London. London, nd (1920s). 326pp. 8x10 green decorated cloth. G. **$63.00\£35.00** *(ABx 3/89)*

MacLeod, Dawn. Oasis of the North. London, 1963. 2nd ed. 268pp. G in dj. **$18.00\£10.00** *(ABx 3/89)*

MacLeod, Dawn. The Gardener's London. London, 1972. 336pp. Internally F in wrn clth and dj. **$35.00\£19.00** *(BF 2/90)*

MacMillan, Conway. Minnesota Plant Life. St. Paul, 1899. Sm 4to, 568pp. Spine end wear. G. **$8.00\£4.00** *(BLB 7/89)*

Macmillan, H.F. Tropical Gardening and Planting. London, 1925. 3rd ed. [xvi], 594pp. Binder's clth, slght rodent damage to outer blank crnrs. **$31.00\£17.00** *(WW 1/90)*

MacNeal, D.L. Flora of the Upper Cretaceous Woodbine Sand, Denton County, Texas. 1958. (Acad. Nat. Science, Phila., Monogr. 10). F. **$14.00\£8.00** *(PBx 11/89)*

Macneil, Alan & Esther. Garden Lilies. NY, 1946. 1st ed. [xiv], 226pp. VG in tattered dj. **$9.00\£5.00** *(CB 12/89)*

Macoboy, Stirling. Tropical Flowers and Plants. 1974. 1st prntng. 128pp. VG in dj. **$18.00\£10.00** *(CB 12/89)*

Macoun, Dr. W.T. Hedges and Their Uses. Ottawa, 1931. 1st, softbound. 44pp. VG/F. **$4.75\£3.00** *(BLB 7/89)*

MacSelf, A.J. A Concise Dictionary of Gardening. GC, 1940. 544pp. VG/F. **$8.00\£4.00** *(BLB 7/89)*

Macself, A.J. Ferns for Gardens and Greenhouse. 1st ed. 1952. Scarce. VG. **$27.00\£15.00** *(DL 89/90)*

MacSelf, A.J. Hardy Perennials. London, 1950. 1st edn of updated edn, 304pp. G/VG. **$10.50\£6.00** *(BLB 7/89)*

Macself, A.J. Hardy Perennials. London, NY, 1950. 304pp. VG. **$17.00\£9.00** *(CB 1/90)*

MacSelf, A.J. Rose Grower's Treasury. London, 1934. 352pp. Cloth lightly worn else VG. **$15.00\£8.00** *(AB 4/89)*

Macself, A.J. The Amateur's Greenhouse. London, 1974. 5th ed., rev. 335pp. F in dj. **$11.50\£6.00** *(BLB 7/89)*

MacSelf, A.J. The Chrysanthemum Grower's Treasury. London, 1948. Rev ed. 212pp. VG. **$4.50\£3.00** *(BB 7/89)*

MacVeagh, Mrs. Charles. Fountains of Papel Rome. NY, 1915. [xiv], 312pp. Spine faded else VG. **$45.00\£25.00** *(Hrt 12/89)*

Madge, Kathleen. The World of Living Green. London, 1947. 120pp. F in dj. **$7.50\£4.00** *(WH 1/90)*

Maeterlinck, Maurice. Old Fashioned Flowers and Other Open-Air Essays. Toronto, nd (c. 1906). 115pp. Full green calf, spine sunned. 6x7.5. G. **$32.50\£18.00** *(ABx 3/89)*

Maeterlinck, Maurice. Old Fashioned Flowers. NY, 1905. Tall 12mo. 106pp. Decorated cover. VG. **$20.00\£11.00** *(RW 11/86)*

Another copy. London, 1906. 115pp. Spine parts missing, internally loose, worn. All plates present. **$15.00\£8.00** *(AB 2/89)*

Another copy. NY, 1905. 1 plate bnd out of context else about F. **$25.00\£14.00** *(Hrt 12/89)*

Maeterlinck, Maurice. The Intelligence of the Flowers. NY, 1907. 179pp. F in patterned paper on brds, buckrm spine. **$150.00\£83.00** *(BF 2/90)*

Maher, James T. The Twlight of Splendor. Boston, 1975. [xix], 453pp. VG in dj. **$35.00\£19.00** *(Hrt 12/89)*

Mahler, W. Keys to the Vascular Plants of the Black Gap Wildlife Management Area Brewster County, Texas. S. Methodist, 1971. 109pp. Wrs. VG. **$11.00\£6.00** *(GW 8/89)*

Maiden, J.H. The Useful Native Plants of Australia, Including Tasmania. London, 1889. 1st ed. [xii], 696pp. 8vo, green cloth. Rare. Unopened. Near fine. **$115.00\£64.00** *(BB 12/88)*

Maire, Dr. Rene. Les Progres des Connaissance Botniques en Algeria depuis 1830. Paris, 1931. Sm 4to, heavy paper covers. 227pp. VG. **$9.50\£5.00** *(BLB 7/89)*

Makins, F.K. Herbaceous Garden Flora. London, 1957. G. **$18.00\£10.00** *(WW 1/90)*

Malby, Reginald. The Story of My Rock Garden. London, 1913. 3rd ed. 132pp. VG. **$10.00\£6.00** *(RW 11/86)*

Another copy. London, nd (1913). 129pp. G. **$21.50\£12.00** *(ABx 3/89)*

Another copy. Toronto, 1912. 2nd ed., F. **$28.00\£16.00** *(PBx 11/89)*

Mallary, P. & F. A Redoute Treasury. 1968. Folio. F in dj. **$45.00\£22.00** *(PBx 11/89)*

Mangelsdorf & Reeves. Origin of Indian Corn and Its Relatives. Tx. Agri. Exp. St., 1939. 315pp. Paper, light wear. **$38.00\£21.00** *(AB 2/89)*

Mansfield, Louise. An Artist's Herbal. NY, 1937. [xiv], 77pp. 4to. VG in dj. **$15.00\£8.00** *(RW 11/86)*

Another copy. NY, 1937. 76pp. 4to. G in dj. **$45.00\£25.00** *(ABx 3/89)*

Another copy. NY, 1937. [xiv], 76pp. G. **$20.00\£11.00** *(Hrt 12/89)*

Mansfield, T.C. Alpines in Colour and Cultivation. London, 1942. 1st, 277pp. Spine tear repaired, few light pencil marks. Bit loose. G. **$10.50\£6.00** *(BLB 7/89)*

Mansfield, T.C. Roses in Colour and Cultivation. London, 1943. 1st, 263pp. Becoming scarce. VG/F. **$11.75\£7.00** *(BLB 7/89)*

Another copy. London, 1946. Rev. ed. VG. **$18.00\£10.00** *(PBx 11/89)*

Mansfield, T.C. Shrubs in Colour and Cultivation. London, 1946. Rev edn. 267pp. Cover edge fade. G. **$8.00\£4.00** *(BLB 7/89)*

Markham, E. Clematis. London, 1935. Spine faded. G. **$27.00\£15.00** *(WW 1/90)*

Markham, Ernest. Clematis. London, 1939. 2nd ed. 116pp. G in repaired dj. **$21.50\£12.00** *(ABx 3/89)*

Marret, L., L. Capitaine and R. Rarrer. Icones Florae Alpinae Plantarum. Paris, 1911-1919. Series 1 & 2. In 10 prts, ry 8vo, orig wrpprs. G. **$63.00\£35.00** *(WW 1/90)*

Marsden, C. Grow Cacti. London, 1958. 2nd edn. 178pp. VG in dj. **$11.75\£7.00** *(BLB 7/89)*

Marsden-Jones, E. British Knapweeds: A Study in Synthetic Taxonomy. Ray., 1954. 201pp. VG in dj. **$15.00\£8.00** *(GW 8/89)*

Marshall, C. A Plain and Easy Introduction to the Knowledge and Practice of Gardening. London, 1813. 5th ed. [iv], 448pp. New clth, G. **$51.00\£28.00** *(WW 1/90)*

Marshall, Charles. A Plain and Easy Introduction to the Knowledge and Practice of Gardening. 3rd ed. 1800. Later green half calf gilt bndng. 408pp. F. **$145.00\£81.00** *(PBx 11/89)*

Marshall, N. The Mushroom Book. Doubleday, 1923. 173pp. Scuffed covers. **$20.00\£11.00** *(GW 12/88)*
Another copy. NY, 1919. 170pp. G. **$11.75\£7.00** *(BLB 7/89)*

Martin, Alexander C. & William D. Barkley. Seed Identification Manual. Berkeley, U. of Calif. Pr., 1961. 221pp. 4to. VG in tattered dj. **$45.00\£25.00** *(BB 12/88)*

Martin, Isa (ed). The Field Club Flora of the Lothians. London & Edinburgh, 1927. 142pp. G. **$27.00\£15.00** *(ABx 3/89)*

Martin, M., P. Chapman & H. Auger. Cacti. London, 1971. 1st, 207pp. As New in dj. **$10.50\£6.00** *(BLB 7/89)*

Martin, W. Keble. Over the Hills. London, 1968. 1st, 173pp. F in dj. **$8.00\£4.00** *(BLB 7/89)*
Another copy. London, 1968. 176pp. VG in dj. **$20.00\£11.00** *(JI 7/88)*

Martin, W.K. The Concise British Flora. London, 1965. Ry 8vo. Trifle used. **$21.00\£12.00** *(WW 1/90)*

Martineau, Alice. The Herbaceous Garden. London, 1913. 2nd imp. Lge 8vo. [xx], 298pp. VG. **$25.00\£14.00** *(RW 11/86)*

Martineau, P. Gardening in Sunny Lands – The Riviera, California, Australia. 1924. Scarce. Well preservd cpy. **$45.00\£25.00** *(DL 89/90)*

Marzell, H. ed. Woerterbuch der Deutschen Pflanzennamen. Parts 1-15. (Abelia to Hypericum). 1937-1958. As issued. **$135.00\£75.00** *(PBx 11/89)*

Masefield, Charles. Native Plants. Wellington, 1958. Reprint, 64pp. Endpaper clipped else G. **$5.50\£3.00** *(BLB 7/89)*

Masefield, G.B. et al. Oxford Book of Food Plants. London, 1969. G. **$27.00\£15.00** *(WW 1/90)*

Massee, George. European Fungus Flora Agaricaceae. London, 1902. 1st ed. [vi], 274pp. Small tear in page 1, sl fading to spine else F. **$30.00\£17.00** *(JI 12/88)*

Massie, Susanne Williams and Francis Archer Christian. Homes and Gardens in Old Virginia. Richmond, 1932. [xviii], 367pp. VG. **$15.00\£8.00** *(RW 6/89)*

Massingham, Betty. Flower Arranging in Colour. London, 1970. 2nd imp. 301pp. VG in dj. **$15.00\£8.00** *(RW 6/89)*

Massingham, Betty. Garden Flowers for the House. London, 1971. 226pp. 4to. G. **$28.00\£16.00** *(ABx 3/89)*

Massingham, Betty. Miss Jekyll, Portrait of a Great Gardener. Newton Abbot: 1973. Rev ed. F in dj. **$25.00\£14.00** *(BF 2/90)*

Massingham, Betty. Miss Jekyll. London, 1973. Rev. ed. 195pp. G in dj. **$25.00\£14.00** *(ABx 3/89)*

Masson, Georgina. Italian Gardens. NY, 1961. 300pp. Folio. F in tan clth, edges slightly brwnd. **$90.00\£50.00** *(BF 2/90)*

Mathew, B. Dwarf Bulbs. NY, 1973. As new in dj. **$22.00\£12.50 (DL 89/90)**

Mathews, F. Schuyler. Familiar Flowers of Field and Garden. NY, 1895. 308pp. VG in slightly rbbd pict clth. **$20.00\£11.00** *(BF 2/90)*

Mathews, F. Schuyler. Familiar Flowers of Field and Garden. NY, 1903. [ix], 308pp. VG. **$15.00\£8.00** *(RW 11/86)*

Mathews, F. Schuyler. Familiar Trees and Their Leaves. NY, 1901. [xv], 320pp. Slightly shaken. Wear at ends of spine. **$12.50\£7.00** *(BB 12/88)*

Mathews, F.S. Field Book of American Wildflowers. NY, 1902. Original edn. 552pp. Pencil notes else G. **$7.00\£4.00** *(BLB 7/89)*

Matschat, C. How to Make a Garden. Boston, 1937. 91pp. VG. **$12.00\£7.00** *(BF 2/90)*

Matschat, Cecil H. The Garden Calendar. Boston, 1936. 12mo. 118pp. Ex-Darien Garden Club libry, with their bkplte else VG in lghtly spttd clth. **$12.00\£7.00** *(BF 2/90)*

Matschat, Cecile H. Mexican Plants for American Gardens. Boston, 1935. 269pp. VG. **$25.00\£14.00** *(BF 2/90)*

Matschatt, Cecile. American Wild Flowers. NY, 1940. 4to. VG in dj. **$15.00\£8.00** *(WH 1/90)*

Matson, Ruth. Gardening for Gourmets. GC, 1958. 262pp. VG/F. **$5.75\£3.00** *(BLB 7/89)*

Mawe, Thomas & John Abercrombie. Every Man His Own Gardener. London, 1787. 11th ed. 12mo. [4], 616pp, [19]. Pin pricks in pages front and back else a nice copy in 1/4 leather on marbled brds. **$65.00\£36.00** *(Hrt 12/89)*

Mawson, Thomas H. The Art & Craft of Garden Making. London and NY, 1912. 4th ed. Folio, gilt decorated green cloth. T.e.g., 404pp. Hinges cracked. Shelf worn. **$150.00\£83.00** *(MD 7/89)*

Another copy. London and NY, 1912. 4th ed. 404pp. Spine slightly darkened else G. **$171.00\£95.00** *(ABx 3/89)*

Maxwell, D. Fyfe & P.S. Patrick. The English Heather Garden. London, 1966. 1st ed. 184pp. G in dj. **$21.50\£12.00** *(ABx 3/89)*

Maxwell, H. Trees. Glasgow, 1915. [xvi], 236pp. Sm 4to. Trifle used, a little fxng. **$36.00\£20.00** *(WW 1/90)*

Maxwell, Sir Herbert. Flowers A Garden Notebook. Glasgow, 1923. Deluxe ed. On handmade paper, limited to 300 copies. This #30. 250pp. Quarter calf, 4to. VG. **$90.00\£50.00** *(ABx 3/89)*

Maxwell, Sir Herbert. Scottish Gardens. London, 1908. 2nd imp. 252pp. 8x10 decorative cloth. G. **$63.00\£35.00** *(ABx 3/89)*

Maxwell, Sir Herbert. Scottish Gardens. NY and London, 1911. New ed. 206pp. Fxd in rather wrn clth. **$45.00\£25.00** *(BF 2/90)*

May, W.J. Greenhouse Management for Amateurs. London, nd (c. 1884). 3rd ed. VG. **$20.00\£11.00** *(RW 11/86)*

Maynard, S.T. The Practical Fruit Grower. NY, 1906. 12mo. 128pp. **$20.00\£11.00** *(RW 6/89)*

Maynard, Samuel. Landscape Gardening As Applied to Home Decoration. NY, 1899. 1st, 338pp. Spine fade, hinges cracked. Light cover soil. G. **$8.00\£4.00** *(BLB 7/89)*

Maynard, Samuel. Landscape Gardening. NY, 1914. 1st ed. [xvi], 338pp. VG. **$15.00\£8.00** *(RW 11/86)*

McAlpine, D. Fungus Diseases of Citrus Trees in Australia. Melbourne, 1899. 132pp. Wrpprs. Used. **$36.00\£20.00** *(WW 1/90)*

McCalla, W.C. Wild Flowers of Western Canada. Toronto, 1920. 1st, 132pp. Quite scarce. VG. **$11.50\£6.00** *(BLB 7/89)*

McCann, Charles. 100 Beautiful Trees of India. Bombay, 1959. 2nd ed. 168pp. VG in dj. **$25.00\£14.00** *(Hrt 12/89)*

McCauley, Lena May. The Joy of Gardens. NY, 1911. [x], 239pp. 7 of garden plans. VG. **$25.00\£14.00** *(Hrt 12/89)*

McCauley, Lena. The Joy of Gardens. Chicago, 1911. [x], 239 + 7pp of plans, 48 photographic plates. VG. **$10.00\£6.00** *(RW 11/86)*

McClintock, Elizabeth. A Monograph on the Genus Hydrangea. SF, 1957. CAS proceedings, vol xxix, no. 5. 147-256pp. Glossy paper in gray wrappers. VG. **$15.50\£8.00** *(BB 12/88)*

McClure, F.A. The Bamboos. Cambridge, 1966. 347pp. F in dj. **$45.00\£25.00** *(WH 1/90)*

McCollum, William. Vines and How to Grow Them. NY, 1914. 12mo. 319pp. VG. **$7.50\£4.00** *(RW 11/86)*

McCully, Anderson. American Alpines in the Garden. NY, 1931. 251pp. VG. **$18.00\£10.00** *(AB 4/89)*
Another copy. NY, 1931. 1st ed. [viii], 251pp. VG in tattered dj. **$20.00\£11.00** *(CB 12/89)*

McCurdy, D., W. Spangenberg & C. Doty. How to Choose Your Trees. Carbondale, 1972. 4to, 245pp. As New in dj. **$6.50\£4.00** *(BLB 7/89)*

McDonald, Elvin. The Complete Book of Gardening under Lights. GC, 1965. 215pp. F in dj. **$6.00\£3.00** *(BLB 7/89)*

McDonald, W.C. Diseases of Field Crops in the Prairie Provinces. Ottawa, 1967. Rev edn, softbound. 100pp. Light corner bump. VG. **$4.50\£3.00** *(BLB 7/89)*

McDougal, W.B. Plant Ecology. Phila, 1949. 4th ed. rev. 234pp. F. **$10.00\£6.00** *(WH 1/90)*

McDougall, W.B. & Herma A. Baggley. Plants of Yellowstone Park. GPO, 1936. Wrps. 160pp. Spine chppd else VG. **$10.00\£6.00** *(CB 1/90)*

McDougall, W.B. Grand Canyon Wild Flowers. Flagstaff, 1964. 1st ed. [x], 262pp. VG. **$20.00\£11.00** *(JI 4/89)*

McEachern, Neil. The Villa Taranto. London, 1954. 88pp. G. **$32.50\£18.00** *(ABx 3/89)*

McElhanney, T.A. et al. Canadian Woods. Ottawa, 1935. 1st ed. [xvi], 348pp. Sm 4to. VG. **$45.00\£25.00** *(JI 12/88)*

McFadden, Dorothy. Gardens of Europe. NY, 1970. 4to. 198pp. VG in dj. **$20.00\£11.00** *(RW 11/86)*
Another copy. NY, 1970. 198pp. Back brd tender else VG in rubbed dj. **$35.00\£19.00** *(Hrt 12/89)*

McFarland, Horace. Getting Acquainted with Trees. NY, 1904. [xi], 241pp. VG. **$15.00\£8.00** *(RW 6/89)*

McFarland, Horace. Roses of the World. Boston, 1947. 3rd ed. [xxiv], 296pp. VG in tattered dj. **$14.00\£8.00** *(CB 1/90)*

McFarland, J. et al. Garden Bulbs in Color. 296pp. Author. 1945. VG. **$11.00\£6.00** *(GW 12/88)*

McFarland, J. Horace. My Growing Garden. NY, 1915. 216pp. G. **$25.00\£14.00** *(Hrt 12/89)*

McFarland, J. Horace. Roses of the World in Color. Boston, NY, 1938. 2nd ed. [xxiv], 296pp. VG. **$15.00\£8.00** *(CB 12/89)*

McFarland, J. Horace. The Rose in America. NY, 1923. 1st ed. [x], 233pp. G. **$10.50\£6.00** *(CB 12/89)*

McGredy, Sam & S. Jennett. A Family of Roses. NY, 1971. 1st US. 168pp. G in dj. **$7.00\£4.00** *(BLB 7/89)*

McGuire, Frances. Gardens of Italy. NY, 1964. 223pp. VG in dj. **$18.00\£10.00** *(AB 4/89)*
Another copy. Barrows, 1964. 223pp. VG in dj. **$12.50\£7.00** *(GW 12/88)*

McHarg, Ian. Design with Nature. Garden City, for Amer Museum of Natural History, 1969. 1st ed. [viii], 197pp. VG. **$70.00\£39.00** *(Hrt 12/89)*

McHarg, Ian. Design with Nature. NY, 1969. 1st ed. Sq. 4to. [viii], 197pp. VG in worn dj. **$15.00\£8.00** *(RW 11/86)*

McHatton, Thomas H. Armchair Gardening: Some of the Spirit, Philosophy and Psychology of the Art of Gardening. Athens, GA, 1947. 130pp. VG in wrn dj. **$12.00\£7.00** *(BF 2/90)*

McIlvaine, C. & R. MacAdam. One Thousand American Fungi. W. Glover, 1973. 729pp. G in dj. **$11.75\£7.00** *(BLB 7/89)*

McKay, C.D. The French Garden. London, 1908. 62pp. G. **$7.25\£4.00** *(ABx 3/89)*

McKelvey, S. Botanical Exploration of the Trans-Mississippi West 1790-1850. 1144pp. Harvard, 1955. Scarce. VG. **$200.00\£111.00** *(GW 8/89)*

McKenny, Margaret. Birds in the Garden and How to Attract Them. NY, 1939. 1st, sm 4to. 349pp. Sm edge nick. G. **$8.00\£4.00** *(BLB 7/89)*

McKenny, Margaret. The Wild Garden. Garden City, 1936. [xi], 123pp. Front edge damaged else VG. **$9.00\£5.00** *(CB 12/89)*

McKenny, Margaret. Trees of the Countryside. NY, 1942. 1st. 64pp. light cover soiling & wear. G/VG. **$7.00\£4.00** *(BLB 7/89)*

McLaren, John. Gardening in California. Landscape and Flower. S.F., 1924. 3rd ed. [xii], 395pp. Inscrbd by the printer. Pg edges untrimmed, 2pp have sm chunks mssng from poor cutting. Gilt pict cvr. some shelfwr else VG. **$50.00\£28.00** *(CB 1/90)*

McMinn, Howard & Evelyn Maino. An Illustrated Manual of Pacific Coast Trees. Berkeley, 1969. 2nd ed. 409pp. G/VG. **$9.50\£5.00** *(BLB 7/89)*

McMinn, Howard. An Illustrated Manual of California Shrubs. SF, 1939. [xi], 689pp. Name on dedic pg, tips bmpd else VG. **$27.00\£15.00** *(CB 1/90)*

McMinn, Howard. An Illustrated Manual of Californial Shrubs. SF, 1939. 1st ed. [xi], 689pp. VG. **$39.00\£22.00** *(BB 12/88)*

McPhee, John. Oranges. NY, 1967. 1st ed. 149pp. F in dj. **$50.00\£28.00** *(Hrt 12/89)*
Another copy. NY, 1967. 1st prntng. 149pp. VG in dj. **$50.00\£28.00** *(CB 1/90)*

McVaugh, Rogers. Edward Palmer, Plant Explorer of the American West. Norman, 1956. 1st ed. 430pp. F in dj. **$20.00\£11.00** *(WH 1/90)*

McVean, Donald & Derek Ratcliffe. Plant Communities of the Scottish Highlands. HMSO, 1962. 445pp. 2 maps in back pocket. G. **$72.00\£40.00** *(ABx 3/89)*

McVickar, Dr. M.H. Using Commercial Fertilizers. Danville, 1962. 2nd ed. 286pp. VG/F. **$5.75\£3.00** *(BLB 7/89)*

McWatt, J. The Primulas of Europe. London, 1923. [xvi], 208pp. Cr 8vo, brds, bndng trifle soiled, inscrip on endpaper. **$72.00\£40.00** *(WW 1/90)*

Meachem, W.L. An Easy Guide to African Violets. NY, 1956. 1st, 64pp. VG/F in worn dj. **$5.75\£3.00** *(BLB 7/89)*

Mee, Margaret. Flowers of the Brazilian Forests. London, 1968. Ltd. ed. #400/500. Folio. Bnd by Zaehnsdorf, hlf morocco on mrbld brds; in slip case. As new. **$1300.00\£720.00** *(Hrt 12/89)*

Meggers, B. et al. Tropical Forest Ecosystems in Africa & S. America: A Comparative Review. Smithsonian, 1973. 350pp. VG in dj. **$20.00\£11.00** *(GW 8/89)*

Meier, W.H.D. School and Home Gardens. Boston, 1913. [v], 319pp. Spine fade else VG. **$18.00\£10.00** *(Hrt 12/89)*

Menninger, Dr. E.A. Fantastic Trees. NY, 1968. 2nd prntng. Lg 8vo, 304pp. VG/F. **$11.75\£7.00** *(BLB 7/89)*

Mentz, A. & C.H. Ostenfeld. Billeder af Nordens Flora. Copenhagen & Stockholm, 1901-1903. 3 vols. Vol. I: [x], 268pp. Vol. II: [xi], 269-519 col plts. Vol. III: [ix], 469pp. Wrn, hlf lthr marbld brds. Plts & text VG. **$625.00\£347.00** *(Hrt 12/89)*

Mercer, F.A. (ed). Gardens & Gardening. London, 1932-1940. Run of 9 vols. Clth edges dmp stn on several vols. Some fxng f.f.e.p.s. else quite G in worn djs. **$270.00\£150.00** *(Hrt 12/89)*
Another copy. London and NY, 1934. 128pp. Green clth, water stn bck cvr. Contents VG. **$20.00\£11.00** *(CB 1/90)*

Mercer, F.A. (ed). Gardens and Gardening. London, 1932. [v], 122pp. VG. **$25.00\£14.00** *(RW 6/89)*

Mercer, F.A. (ed). Gardens and Gardening. London, 1934. 128pp. VG. **$25.00\£14.00** *(RW 6/89)*

Mercer, F.A. & Roy Hay (eds). Gardens and Gardening. 1950. London, 1950. 172pp. VG in dj. **$15.00\£8.00** *(RW 11/86)*
Another copy. London, 1950. 172pp. G. **$16.00\£9.00** *(Hrt 12/89)*

Mercer, F.A. & Roy Hay. Gardens and Gardening. Vol. 3. London, 1952. 142pp. VG in dj. **$12.50\£7.00** *(RW 6/89)*

Mercer, Frank & Roy Hay. Hardy Plants. London, 1952. 142pp. VG in torn dj. **$8.50\£5.00** *(BLB 7/89)*

Merriman, P.R. Flora of Richmond & Vicinity. Richmond, VA, 1930. Pbnd. 353pp. G. **$28.00\£16.00** *(PBx 11/89)*

Merriman, P.R. Flora of Richmond and Vicinity. Richmond, 1930. 1st, 353pp. Rebound in cloth. G. **$10.50\£6.00** *(BLB 7/89)*

Messegue, Maurice. Of Men and Plants. NY, 1973. 1st, 327pp. As New in dj. **$4.50\£3.00** *(BLB 7/89)*

Metzler, Rejean. Forever Flowers. NY, 1972. 157pp. G in dj. **$7.00\£4.00** *(BLB 7/89)*

Meyer, F.W. Rock and Water Gardens. London, nd (c. 1910). [xii], 227pp + 14pp ads. Partially uncut. Spine faded, else VG. **$20.00\£11.00** *(RW 11/86)*

Meyer, Joe. Herbalist. Hammond, 1934. 400pp. VG. **$15.00\£8.00** *(AB 4/89)*
Another copy. Hammond, 1960. 304pp. As New in dj. **$12.00\£7.00** *(AB 4/89)*
Another copy. Np, 1975. 10th prntng, narrow 24mo. 304pp. Cover rubbed. VG in dj. **$7.00\£4.00** *(BLB 7/89)*

Meyer, Joseph. The Herbalist. Hammond, 1934. Narrow 12mo. 400pp. Extremities wrn, some soiling else VG. **$21.00\£12.00** *(CB 1/90)*

Micheli, M. Le Jardin du Crest; Notes sur les Végétaux Cultivés en Plein Air au Chateau du Crest. Genève, 1896. Autographed dedication to his friend Vilmorin. Covers frayed else VG. **$75.00\£42.00** *(PBx 11/89)*

Michelia. Commentarium Mycologiae Italicae. Amsterdam, 1969. 1301pp. 2 vols. All pblshd. G. **$72.00\£40.00** *(WW 1/90)*

Middleton, C.H. More Gardening Tales. London, 1938. 12mo. 256pp. Ownr's sig, slght fxng to e.p. Else VG. **$15.00\£8.00** *(BF 2/90)*

Middleton, C.H. Mr. Middleton Advises/More Gardening Talks. London, 1941. VG. **$12.00\£7.00** *(BF 2/90)*

Middleton, C.H. Your Garden in War-Time. London, 1941. 12mo. 200pp. Ownr's name on e.p. else VG. **$15.00\£8.00** *(BF 2/90)*

Migula, W. Kryptogamen - Flora von Deutschland. Gera, 1911-1912. Band 3, Pilze, Teil 2, Basidiomycetes. 2 vols. New cloth. **$180.00\£100.00** *(WW 1/90)*

Miklos, J. von & E. Flore. The History, the Beauty, the Riches of the Gardener's World. NY, 1968. Brds, G. **$36.00\£20.00** *(WW 1/90)*

Milais, J.G. Rhododendrons. . . The Genus Rhododendron Including Azaleas and the Various Hybrids. 2nd series. 1924. 265pp. #498 of 550 copies. Corners sl bumped, few minor cover imperfections else sound and clean. **$275.00\£153.00** *(PBx 11/89)*

Miles, Bebe. Bluebells and Bittersweet. NY, 1970. 168pp. VG. **$8.50\£5.00** *(BLB 7/89)*

Mill, Anna Jean. John Mill's Boyhood Visit to France. Toronto, 1960. [xxxi], 133pp. VG in dj. **$15.00\£8.00** *(Hrt 12/89)*

Miller, Ellen & Margaret C. Whiting. Wild Flowers of the Northeastern States. NY, 1895. Lge 4to. [xi], 622pp. Decorated cover. Fine. Scarce. **$45.00\£25.00** *(RW 11/86)*

Miller, Julian. A Monograph of the World Species of Hypoxylon. Athens, 1961. 158pp. F. **$10.00\£6.00** *(WH 1/90)*

Miller, Orson. Mushrooms of North America. NY, 1972. Sm 4to, 360pp. VG in dj. **$15.25\£8.00** *(BLB 7/89)*

Miller, Philip. The Gardener's Dictionary: Containing the Methods of Cultivating and Improving the Kitchen. . . Physic Garden. . . According to the Practice of the Most Experienced Gardeners of the Present Age, etc. 1731. 1st ed. Folio. Orig clf; unusually well preservd. **$540.00\£300.00** *(DL 89/90)*

Miller, Robert. Root Anatomy and Morphology. Hamden, 1974. 1st, sm 4to. 271pp. As New in dj. **$9.25\£5.00** *(BLB 7/89)*

Miller, Wilhelm. What England Can Teach Us About Gardening. NY, 1911. 3rd prntng. 359pp. G. **$81.00\£45.00** *(ABx 3/89)*

Millett, Mervyn. Australian Eucalypts. Melbourne, 1969. 1st, heavy cardboard covers. 113pp. Minimal cover defects else G. **$8.00\£4.00** *(BLB 7/89)*

Mills, Carolyn & David Wilson. Flower Arranging and Flower Shop Operation. LA, 1959. 2 vols, 4tos, softbound. 326 + 197pp. Some cover wear else G. Set, **$10.50\£6.00** *(BLB 7/89)*

Mills, John, (trans). (Duhamel du Monceau). A Practical Treatise of Husbandry. London, 1759. [xxiv], 492pp, (7), of index (1). Latter period 1/2 lthr on mrbld brds. VG. **$400.00\£222.00** *(Hrt 12/89)*

Millspaugh, C.F. Flora of the Sand Keys of Florida. Chicago, 1907. 1st, softbound. 54pp. About Fine. **$8.00\£4.00** *(BLB 7/89)*

Milman, Helen (Mrs. Caldwell Crofton). My Kalendar of Country Delights. London, 1903. T.e.g. [vii], 368pp. Dark spine, covers smudged else VG. **$15.00\£8.00** *(Hrt 12/89)*

Milne, Lorus & Margery. Living Plants of the World. NY, 1967. Rev. edn., 4to. 336pp. G in dj. **$17.75\£10.00** *(BLB 7/89)*

Milne-Holme, Mary P. Stray Leaves from a Border Garden. London, 1902. 340pp. Ownr's bkplte, e.p.s. fxed else VG. **$25.00\£14.00** *(BF 2/90)*

Milner, Mary. The Garden, the Grove and the Field: A Garland of the Months. Bath, nd (1853?). 242pp. Decorated cloth. G. **$27.00\£15.00** *(ABx 3/89)*

Milton, John. The Rose Question and Answer Book. NY, 1965. 192pp. As New in dj. **$4.50\£3.00** *(BLB 7/89)*

Missouri Botanical Gardens. Second Annual Report. 1890. 53 pls. VG. **$15.00\£8.00** *(GW 8/89)*

Missouri Botanical Gardens. Sixth Annual Report. 1895. 62 pls. VG. **$15.00\£8.00** *(GW 8/89)*

Missouri Botanical Gardens. Tenth Annual Report. 1899. 62 pls. VG. **$15.00\£8.00** *(GW 8/89)*

Mitchell, Donald. My Farm of Edgewood, A Country Book. NY, 1864. 319pp. Binding worn, contents clean but loose. **$20.00\£11.00** *(AB 2/89)*

Mitchell, Donald. Rural Studies with Hints for Country Places. NY, 1867. 295pp. Green cloth, light wear, tips bumped, else clean and tight. 1st ed. Scarce. **$50.00\£28.00** *(AB 4/89)*

Mitchell, Donald. Wet Days at Edgewood. NY, 1894. [vii], 324pp. VG. **$20.00\£11.00** *(RW 6/89)*

Mittwer, Henry. The Art of Chabana. Rutland, VT, 1974. 1st ed. 144pp. 8vo. VG in dj. **$15.00\£8.00** *(JI 12/88)*

Moeliono, B.M. Cauline or Carpellary Placentation among Dicotyledons. 1970. 292pp. 2 vols. F. **$40.00\£22.00** *(PBx 11/89)*

Mohlenbrock, R.H. Illustrated Flora of Illinois. Carbondale. SIUP, 1970. 272pp. As New in dj. **$9.25\£5.00** *(BLB 7/89)*

Monteath, Robert. The Forester's Guide and Profitable Planter. London, 1836. 3rd ed. [xi], 500pp. Half calf, marbled end-papers, edges. Boards, G. **$75.00\£42.00** *(WH 1/90)*

Montgomery, E.G. Productive Farm Crops. Philadelphia, 1918. Rev 2nd edn. 507pp. Top edge smoke darkened else G/VG. **$5.75\£3.00** *(BLB 7/89)*

Montgomery, F.H. Native Wild Plants of Eastern Canada and the Adjacent Northeastern United States. Toronto, 1962. 1st ed. [xxxviii], 194pp. VG in worn dj. **$25.00\£14.00** *(JI 4/89)*

(Montreal Botanical Garden Publ.) Memoirs #1-54. 1940-1960. As issued. Lacks nos. 5, 28-46. G. **$20.00\£11.00** *(PBx 11/89)*

Moore, H.E. Jr. African Violets, Gloxinias and Their Relatives. NY, 1957. 1st, 323pp. VG in dj. **$11.50\£6.00** *(BLB 7/89)*

Moore, T. Nature Printed British Ferns. London, 1859-1860. 2 vols. 8vo. Hlf morocco. Some slt fxing as usual, preliminaries of vol 2 trifle stned. **$270.00\£150.00** *(WW 1/90)*

Moore, T. Nature Printed British Ferns; Being Figures and Descriptions of the Species and Variations of Ferns Found in the United Kingdom. 1859. 2 vols. Ry 8vo. Rebnd in 1/2 green morocco. **$360.00\£200.00** *(DL 89/90)*

Moore, Thomas & Jackman, George. The Clematis As a Garden Flower. Woking Nursery, 1877. New and revised ed. 134pp. Decorated cloth, damp mark on back cover affecting last few pages of text. Slightly loose. Rare. G. **$162.00\£90.00** *(ABx 3/89)*

Moorehead, Alan. Darwin and the Beagle. NY, 1970. 6th ed. 280pp. VG in dj. **$25.00\£14.00** *(Hrt 12/89)*

Moorehead, Alan. The Fatal Impact. NY, 1966. 1st ed. 230pp. VG in dj. **$36.00\£20.00** *(Hrt 12/89)*

Moorehead, Alan. The White Nile. NY, 1960. 1st ed. 385pp. About F in dj. **$36.00\£20.00** *(Hrt 12/89)*

Moquin-Tandon, A. Elements de Teratologie Vegetale. Paris, 1841. [xiii], 403pp. Rather badly waterstained but a very sound copy. **$45.00\£25.00** *(WW 1/90)*

Morcombe, M.K. Australia's Western Wildflowers. Perth, 1968. 4to. 112pp. G in dj. **$12.50\£7.00** *(WH 1/90)*

Moreland, Wallace (ed). Practical Guide to Successful Farming. GC, 1943. 1001pp. About F. **$8.00\£4.00** *(BLB 7/89)*

Moreton, C. Oscar. The Auricula. London, 1964. 90pp. 12x6 card covers. Fine in dj. **$75.50\£42.00** *(ABx 3/89)*

Moreton, Oscar C. Old Carnations and Pinks. London, 1955. [xi], 51pp. Sm nick upper lft crnr of dec brds else VG in dj. **$90.00\£50.00** *(Hrt 12/89)*
Another copy. Gen. ed., 1955. (Ltd to 3000 cpies). VG in dj. **$99.00\£55.00** *(DL 89/90)*

Morin, P. Remarques Necessaires Pour la Culture des Fleurs. Paris, 1694. [xiv], 171 pp. Sm 8vo. Comtemp brds, worn. **$135\£75.00** *(WW 1/90)*

Morley, Brian D. Wild Flowers of the World. NY, 1970. 432 pp. Sm folio. VG in trn, reprd dj. **$22.00\£12.00** *(CB 1/90)*

Morley, Margaret. A Few Familiar Flowers. Boston, 1897. 274pp. VG/F. **$5.75\£3.00** *(BLB 7/89)*

Morley, Margaret. Flowers and Their Friends. Boston, 1897. 255pp. Cover decorations faded, lightly stained. G/VG. **$5.00\£3.00** *(BLB 7/89)*

Morris, F. & E. A. Eames. Our Wild Orchids. NY, 1929. 464 pp. Oldr pict dj. VG. **$75.00\£42.00** *(Hrt 12/89)*

Morrow, Dr. James. A Scientist with Perry in Japan, The Journal of Dr. James Morrow. Chapel Hill, 1947. [xxvi], 307 pp. VG. **$50.00\£28.00** *(Hrt 12/89)*

Morse, Edward S. Japanese Homes and Their Surroundings. NY, 1889. 372pp. Dec blue clth, t.e.g. Spine & bck cvr chpped, 1 leaf stained and scrtchd. Shelfworn. **$65.00\£36.00** *(MD 2/90)*

Morse, H.K. Garden Easily. NY, 1942. 1st, 208pp. VG/F. **$8.00\£4.00** *(BLB 7/89)*

Morton, J.K. West African Lilies and Orchids. London, 1961. 1st, 71pp. Pictorial brds, F. **$8.00\£4.00** *(BLB 7/89)*

Morton, J.W. 250 Beautiful Flowers and How to Grow Them. London, ca 1940. 192pp. VG. **$5.75\£3.00** *(BLB 7/89)*

Morton, Julia F. Folk Remedies of the Low Country. Miami, 1974. 176pp. VG in dj. **$20.00\£11.00** *(Hrt 12/89)*

Morwood, William. Traveler in a Vanished Landscape. London, 1973. 244 pp. As New in dj. **$35.00\£19.00** *(Hrt 12/89)*

Morwood, William. Traveller in a Vanished Landscape. London, 1973. 244pp. G in dj. **$25.25\£14.00** *(ABx 3/89)*
Another copy. 1973. 244pp. VG in dj. **$15.00\£8.00** *(GW 8/89)*
Another copy. London, 1973. 1st ed. [xii], 244pp. VG in dj. **$25.00\£14.00** *(JI 12/88)*

Moss, E.H. Flora of Alberta. Toronto. U. of Toronto Press, 1959. 1st ed. [viii], 548pp. VG. **$35.00\£19.00** *(JI 4/89)*

Mossman, Keith. The Pip Book. London, 1973. 117pp. G in dj. **$7.00\£4.00** *(BLB 7/89)*

Moulen, F. Orchids in Australia. Lausanne, 1958. G. **$32.00\£18.00** *(WW 1/90)*

Muenscher, W. C. Poisonous Plants of the United States. 1939. Orig. ed. VG. **$45.00\£25.00** *(DL 89/90)*

Muir, John. Our National Parks. Boston, 1916. 400 pp. About F in older dj. **$36.00\£20.00** *(Hrt 12/89)*

Muirhead, Desmond. Palms. Globe, AZ, 1961. 140pp. Softbound. **$18.00\£10.00** *(BB 12/88)*

Mullgardt, Louis Christian. The Architecture and Landscape Gardening of the Exposition. SF, 1915. 2nd ed. [x], 202 pp. G. **$40.00\£22.00** *(Hrt 12/89)*

Munz, Philip A. A Manual of Southern California Botany. Claremont, 1935. [xxxix], 642pp. Red cloth. Scarce. G. **$44.00\£24.00** *(BB 12/88)*

Munz, Philip. California Desert Wildflowers. Berkeley, 1962. 1st ed. 122pp. F. **$12.50\£7.00** *(WH 1/90)*

Murray, A. The Northern Flora. Edinburgh, 1836. Prt. 1. [xvii], 150pp, [xvi]. Orig. prntnd brds, clth bck. G. **$90.00\£50.00** *(WW 1/90)*

Murray, Pauline. Planning and Planting the Home Garden. NY, 1932. 1st prntng. [xiv], 412pp. VG. **$15.00\£8.00** *(RW 11/86)*

Murrill, Dr. W.A. Familiar Trees. Gainesville, 1946. Lg 8vo, 174pp. Softbnd, autographed. G. **$7.00\£4.00** *(BLB 7/89)*

Myrick, H. Book of Corn. NY, 1904. 2nd rev. 372pp. Cloth worn, contents VG. **$18.00\£10.00** *(AB 2/89)*

N

Nairne, A.K. The Flowering Plants of Western India. London, 1894. [xlvii], 401pp. Inscrp, stmp on hlf-title, annotations on rear e.p. **$36.00\£20.00** *(WW 1/90)*

Nash, Elizabeth T. One Hundred and One Legends of Flowers. Boston, 1927. 340pp. Spine lettering dull else VG. **$20.00\£11.00** *(BF 2/90)*

Nash, J. English Garden Flowers. London, 1948. 30pp. Brds, G. **$27.00\£15.00** *(WW 1/90)*

Nash, John. English Garden Flowers. London, 1948. 32pp. G in slghtly worn dj. **$39.50\£22.00** *(ABx 3/89)*

National Trust of Scotland. Ben Lawers and Its Alpine Flowers. 1972. 87pp. **$11.00\£6.00** *(GW 12/88)*

Neal, M.C. In Gardens of Hawaii. Honolulu, 1965. New ed. G. **$36.00\£20.00** *(WW 1/90)*

Neame, J. Armstrong. Among the Meadow and Alpine Flowers of Northern Italy. London, 1937. 1st ed. 192pp. Blue cloth mottled else VG in dj. **$54.00\£30.00** *(ABx 3/89)*

Nehrling, Arno & Irene. Peonies Outdoors and In. NY, 1960. 1st ed. 288pp. G in dj. **$14.50\£8.00** *(ABx 3/89)*

Nehrling, Arno & Irene. The Picture Book of Annuals. NY, 1966. 1st, 288pp. VG in dj. **$5.75\£3.00** *(BLB 7/89)*

Neish, R. A World in a Garden. London, 1899. 238pp. Decorations by Arthur Rackham. VG. **$20.00\£11.00** *(RW 6/89)*

Nelmes, E. & W. Cuthbertson. Curtis' Botanical Magazine Dedications, 1827-1927. 1931. Comp. cpy from W. Cuthbertson to R.L. Harrow. Becoming scarce. In a well-preservd state. **$99.00/£55.00** *(DL 89/90)*

Nelson, R.H. (ed). Pyrethrum Flowers. Minn., 1975. 3rd ed. 148pp. VG in dj. **$18.00\£10.00** *(Hrt 12/89)*

Nessel, H. Die Barlappgewachse. Jena, 1939. Ry 8vo. Wrpprs. G. **$32.00\£18.00** *(WW 1/90)*

Neuman, J.J. The Polyporaceae of Wisconsin. Wisconsin Geological and Natural History Survey. Bull #xxxiii. Madison, 1914. 1st ed. [iv], 206pp. VG. **$35.00\£19.00** *(JI 12/88)*

New York Botanical Garden. Bulletin of the NYBG, Volume II, 1901-1903. #6,7,8. [viii], 518pp. 1/4 lthr, chppd mrbld brds, gilt stmpd spine. **$75.00\£42.00** *(Hrt 12/89)*

New York Botanical Garden. North American Flora. Publ. by NYBG: Rosales. 1905-1924. Vol. 22, parts 1-7; Vol. 24, parts 1-4. (As called for; vol. 23 was not publshd). G. **$26.00\£14.00** *(PBx 11/89)*

New York Botanical Garden. North American Flora. Publ. by NYBG: Germaniales & Polygalales. 1907-1924. Vol. 25. Complete in 5 prts. G. **$15.00\£8.00** *(PBx 11/89)*

New York Botanical Garden. North American Flora. Publ. by NYBG: Cardules. 1914-1927. Vols. 33 & 34. Complete in 5 prts. G. **$20.00\£11.00** *(PBx 11/89)*

New York Botanical Gardens. Papers Presented at the Celebration of the 20th Anniversary of the New York Botanical Gardens. 1916. 592pp. Worn chipped wrs. **$20.00\£11.00** *(GW 8/89)*

Newton, J. A Complete Herbal. London, 1798. [xvi], portrait (plte 1) 175 pltes. Mdrn hlf clf, antique style. Small, heavy staining else G. **$324.00\£180.00** *(WW 1/90)*

Nicholls, W.H. Orchids of Australia. Melbourne, 1969. [xxii], 141pp. Folio. Minimal shelf wear on tail edge of rear cover. Well preserved copy in taped dj. **$135.00\£75.00** *(BB 12/88)*

Nichols, B. et al. How Does Your Garden Grow. London, 1935. 1st, 127pp. Light wear, contents VG. **$10.00\£6.00** *(AB 2/89)*

Nichols, Beverley, et al. How Does Your Garden Grow? Garden City. Doubleday Doran, 1935. 1st ed. 139pp. VG in chppd dj. **$25.00\£14.00** *(CB 1/90)*

Nichols, Beverley. A Thatched Roof. NY, 1933. 1st US ed. 294pp. Spine faded. G. **$18.00\£10.00** *(ABx 3/89)*

Nichols, Beverley. A Village in a Valley. Garden City, 1936. 2nd ed. 319pp. VG in chppd, trn dj. **$12.00\£7.00** *(CB 1/90)*

Nichols, Beverley. Down the Garden Path. NY, 1932. Orig edn, 290pp. Minimum cover defects. G. **$8.00\£4.00** *(BLB 7/89)*

Another copy. London, 1932. 1st ed. 290pp. G. **$18.00\£10.00** *(ABx 3/89)*

Another copy. NY, 1932. 1st US. [x], 303pp. Reasonably G copy. **$30.00\£17.00** *(Hrt 12/89)*

Nichols, Beverley. Evensong. London, 1939. Reprnt, 288pp. Minor cover soil else VG. **$4.50\£3.00** *(BLB 7/89)*

Nichols, Beverley. Garden Open Today. London, 1963. 1st, 252pp. VG in dj. **$35.00\£19.00** *(Hrt 12/89)*

Nichols, Beverley. Merry Hall. NY, 1953. 1st Am. ed. VG in dj. **$15.00\£8.00** *(RW 11/86)*
Another copy. NY, 1953. 1st US, 319pp. VG in dj. **$30.00\£17.00** *(Hrt 12/89)*

Nichols, Beverley. Merry Hall. NY, 1953. 1st ed. 319pp. VG in dj. **$17.00\£9.00** *(CB 1/90)*

Nichols, Beverley. Sunlight on the Lawn. London, 1956. 1st ed. 255pp. VG in dj. **$30.00\£17.00** *(Hrt 12/89)*

Nichols, Beverley. The Art of Flower Arrangement. NY, 1967. 239pp. VG in dj. **$50.00\£28.00** *(RW 11/86)*

Nichols, Beverley. Village in the Valley. NY, 1934. 1st Am. ed. 319pp. VG. **$10.00\£6.00** *(RW 11/86)*

Nichols, Rose Standish. English Pleasure Gardens. NY, 1925. [xxii], 324pp. VG in dj. **$55.00\£31.00** *(RW 6/89)*

Nichols, Rose Standish. English Pleasure Gardens. NY, 1925. 2nd prntng. 324pp. VG in dec clth. Inscribed by Nichols. **$70.00\£39.00** *(BF 2/90)*

Nichols, Rose Standish. Spanish & Portuguese Gardens. Boston, 1924. 8vo, green cloth. 304pp. Spine slightly sunned, shelf worn, previous owner's inscription on title. **$45.00\£25.00** *(MD 7/89)*

Nicholson, G. (ed). Illustrated Dictionary of Gardening. Nd, circa 1890-1900. 12 slim vols. Orig green clth. Nice complete set. **$153.00\£85.00** *(DL 89/90)*

Nicol, Walter. The Gardener's Kalendar. Edinburgh, 1814. 3rd ed. [xx], 646pp + 52pp of ads attractively rebound in grey cloth and marbled boards uncut. Prelims slightly marked else a good copy. **$162.00\£90.00** *(ABx 3/89)*

Nicolas, Dr. J.H. A Year in the Rose Garden. GC, 1937. 2nd prntng, 105pp. VG/F in faded dj. **$8.00\£4.00** *(BLB 7/89)*

Nicolas, Dr. J.H. The Rose Manual. NY, 1938. Rev edn. 335pp. Scarce. G/VG. **$9.25\£5.00** *(BLB 7/89)*

Nicolas, J.H. A Rose Odyssey. NY, 1937. 1st ed. 238pp. G. **$18.00\£10.00** *(ABx 3/89)*

Nicolson, Nigel. Great Houses of Britain. NY, 1965. 288pp. VG in dj. **$65.00\£36.00** *(Hrt 12/89)*

Niklitschek, A. Water Lilies and Water Plants. London, 1932. 135pp. G. **$25.25\£14.00** *(ABx 3/89)*

Nisbet, Fred. Growing Better Roses. NY, 1974. 1st, 244pp. VG in worn dj. **$8.00\£4.00** *(BLB 7/89)*

Nisbet, John. Studies in Forestry. Oxford, 1894. 1st ed. [xii], 344pp. Sl cover wear. **$45.00\£25.00** *(JI 12/88)*

Nishikawa, Issotei & Issen. Floral Art of Japan. Tokyo, 1953. 6th prntng. 100pp. VG. **$4.75\£3.00** *(BLB 7/89)*

Nissen, C. Herbals of Five Centuries. Munich, 1958. #53 of 100 cpies. Vol of text. 50 leaves with printed captions. Large folio, unbnd, in orig box. **$2160.00\£1200.00** *(WW 1/90)*

Niwa, Teizo. Chrysanthemums of Japan. London, 1937. 56pp. Rare. G. **$72.00\£40.00** *(ABx 3/89)*

Nixon, Mima, Una Silberrad & Sophie Lyall. Dutch Bulbs and Gardens. London, 1909. 176pp. Decorated cloth. G. **$63.00\£35.00** *(ABx 3/89)*
Another copy. London, 1906. 1st ed. [xii], 176pp. VG. **$30.00\£17.00** *(CB 12/89)*

Noble, Mary & J. Merkel. Gardening in a Small Greenhouse. Princeton, 1958. 2nd prntng. 235pp. VG in dj. **$7.00\£4.00** *(BLB 7/89)*

Noel, M.E. La Vie des Fleurs. Paris, 1860. 4to. Hlf clf, raised bands. A.e.g. end papers else nice and sound. **$40.00\£22.00** *(PBx 11/89)*

Norman, A. Successful Rose Growing. London, 1955. 2nd prntng. 183pp. VG. **$9.25\£5.00** *(BLB 7/89)*

North, Felicity. Gardening in a Cold Climate. London, 1967. 189pp. G in dj. **$21.50\£12.00** *(ABx 3/89)*
Another copy. London, 1967. 189pp. VG in dj. **$25.00\£14.00** *(Hrt 12/89)*

Northen, Rebecca T. Home Orchid Growing. NY, 1974. 3rd ed. 374pp. F in brght, reprd dj. **$23.00\13.00** *(CB 1/90)*

Northend, Mary. Garden Ornaments. NY, 1916. 178pp. G. **$45.00\£25.00** *(ABx 3/89)*

Northend, Mary. Historic Gardens of New England. "The Mentor." Vol 4, #9, June 1916. 12pp + 6 full page half-tones laid in. Wraps. VG. **$20.00\£11.00** *(RW 6/89)*

Norton, Charles Eliot. The Poet Gray as a Naturalist. Boston, 1903. Ltd. ed. 243/500. 67pp. VG. **$120.00\£67.00** *(Hrt 12/89)*

Notcutt, R.C. A Handbook of Flowering Trees and Shrubs. London, 1926. G. **$21.00\£12.00** *(WW 1/90)*

Nuese, Josephine. The Country Garden. NY, 1970. 256pp. VG in dj. **$20.00\£11.00** *(RW 6/89)*
Another copy. NY, 1970. 256pp. VG in dj. **$12.00\£7.00** *(AB 4/89)*

Nussey, Helen G. London Gardens of the Past. London, 1948. 126pp. Scarce. VG in dj. **$55.00\£31.00** *(Hrt 12/89)*

Nuttall, G. Clarke. Beautiful Flowering Shrubs. London, nd (1920). 279pp. Spine faded, generally a little worn and loose. **$39.60\£22.00** *(ABx 3/89)*

Nyman, C.F. Conspectus Florae Europaeae. 3 vols. 1878-82. 858pp. Supplementum I, 1883-84. 859-1046pp, [3]. Supplementum II, 1889-90. 2 prts, 404pp. London, hlf morocco. G. **$81.00\£45.00** *(WW 1/90)*

O

O'Brien, J. Orchids. NY, 1916. 114pp. VG. **$7.00\£4.00** *(BLB 7/89)*

O'Donnell, Thomas C. (ed). A Garden for You. NY, 1946. 199pp. G. **$12.00\£7.00** *(Hrt 12/89)*

Ogden, E. et al. Check-List of the Vascular Plants of Maine. Josselyn, 1966. 71pp. Wrs. VG. **$6.00\£3.00** *(GW 8/89)*

Ogg, Stuart. Delphiniums for Everyone. London, 1961. 1st ed. 96pp. G in dj. **$9.00\£5.00** *(ABx 3/89)*

Okada, Yasuo. Public Lands & Pioneer Farmers. Gage County, NB, 1850-1900. Tokyo, 1971. 188pp. Paper, VG. **$18.00\£10.00** *(AB 2/89)*

Oldham, C.H. The Cultivation of Berried Fruits in Great Britain. London, 1946. 1st, 374pp. VG. **$8.00\£4.00** *(BLB 7/89)*

Oliver, D., W.T. Thiselton-Dyer and D. Prain. Flora of Tropical Africa. London, 1868-1902. Vols. 1-8 in 10 vols. G. **$270.00\£150.00** *(WW 1/90)*

Oliver, G.W. & A.C. Hottes. Plant Culture. NY, 1921. 4th edn. 442pp. VG/F. **$5.75\£3.00** *(BLB 7/89)*

Oneal, Cora M. Gardens and Homes of Mexico. Dallas, 1945. 1st ed. [xiv], 221pp. **$39.00\£22.00** *(BB 12/88)*

Ontario Dept. of Agriculture. Fruits of Ontario. 1914. Bndng loose. Clean ex-lib. Gilt decor clth. **$25.00\£14.00** *(PBx 11/89)*

Organ, John. Gourds. London, 1963. 1st, 188pp. VG/F in dj. **$7.00\£4.00** *(BLB 7/89)*
Another copy. London, 1963. 189pp. G in dj. **$19.75\£11.00** *(ABx 3/89)*

Orr, R.T. & M.C. Wildflowers of Western America. NY, 1974. 1st, lg

8vo. 270pp. F in dj. **$13.00\£7.00** *(BLB 7/89)*

Ortloff, H.S. Perennial Gardens. NY, 1931. 1st, 88pp. VG/F in dj. **$7.00\£4.00** *(BLB 7/89)*

Orton, James. The Andes and the Amazon; or, Across the Continent of South America. NY, 1871. [xxiv], 25-356pp. Except for some gnawing very bottm frnt brd nr spine, VG in gilt lettered clth. **$145.00\£81.00** *(Hrt 12/89)*

Osborn, A. & N. Bagenal (eds). Pruning. London, 1953. 2nd prntng, 160pp. VG in torn dj. **$7.75\£4.00** *(BLB 7/89)*
Another copy. London, 1962. 160pp. G in dj. **$9.00\£5.00** *(ABx 3/89)*

Otis, Charles Herbert. Michigan Trees. Ann Arbor, 1915. [xxxii], 246pp. Ink sig & notes on 1 pg else nr F in glt stmpd clth. **$15.00\£8.00** *(Hrt 12/89)*

Over, W. Flora of South Dakota. SD, 1932. 161pp. Worn wrs. **$11.00\£6.00** *(GW 12/88)*

P

Pack, Charles L. The War Garden Victorious. Phila., 1919. [xiv], 179pp + 2 pamphlets. VG. **$25.00\£14.00** *(RW 11/86)*

Pack, Charles. Trees As Good Citizens. Wash, DC, 1922. 257pp. VG. **$20.00\£11.00** *(RW 6/89)*

Page, Russell. The Education of a Gardener. London, 1971. Reprint. 381pp. G in dj. **$28.75\£16.00** *(ABx 3/89)*

Palaiseul, Jean. Grandmother's Secrets. NY, 1974. 1st US. 292pp. As New in dj. **$7.00\£4.00** *(BLB 7/89)*

Pallas, M.P.S. Voyages de M.P.S. Pallas. Traduits de l'Allemand. Paris 1788-1793. 6 vols. Vol I-V: lg 8vo of text; [xxxii], 773pp, (3) errata; 550pp, (1) errata; 491pp, (1) errata; 722pp, (2) errata, 559pp, (2) errata, respectively. Vol VI: tall 4to huge fldng map, title pg & 124 eng plts; except for moisture stain bottm rght cornr last 20 plts & occas slght mssng edges, a F cpy in early 1/4 lthr on marbld brds; w/armorial Russian bookplt. **$1,350.00\£750.00** *(Hrt 12/89)*

Pammel, Dr. L.H. & C.M. King. Honey Plants of Iowa. Des Moines, 1930. 1192pp. Frt e.p. out, lt damp mks. Hinge reprd. **$20.00\£11.00** *(BLB 7/89)*
Another copy. IA, 1930. 119pp. VG. **$26.00\£14.00** *(GW 12/88)*

Pamplin, W. & A. Irvine. A Botanical Tour in the Highlands of Perthshire. London, 1857. [viii], 78pp. 8vo. Clth, stmp on title pge. G. **$29.00\£16.00** *(WW 1/90)*

Parcher, Emily. Shady Gardens. NY, 1955. 282pp. VG in tattered dj. **$12.00\£7.00** *(CB 1/90)*

Parcher, Emily. Shady Gardens. NY, 1955. Orig edn. 282pp. E.p. out. G. **$7.75\£4.00** *(BLB 7/89)*

Parham, J.W. The Weeds of Fiji. Suva, 1958. [xi], 196pp, [xii-xviii]. Decorated wrappers. VG. **$18.00\£10.00** *(BB 12/88)*

Park, Bertram. Collins Guide to Roses. London, 1956. 1st. 288pp. Owner's inscription on e.p. else VG. **$10.25\£6.00** *(BLB 7/89)*

Park, Bertram. Guide to Roses. Princeton, 1956. 1st US. 288pp. Few pencil marks in pp margs. **$10.50\£6.00** *(BLB 7/89)*

Park, Bertram. The Rose Annual. Croydon, Nat. Rose Soc., GB, 1947 & 1948. 110pp., 180pp. Green cloth, some underlining. VG. **$10.00\£6.00** *(AB 4/89)*

Parker, Eric. Surrey Gardens. London, 1954. 1st ed. 259pp. G in dj. **$25.25\£14.00** *(ABx 3/89)*
Another copy. London, 1954. 259pp. F in dj. **$30.00\£17.00** *(Hrt 12/89)*

Parkhurst, H.E. Trees, Shrubs and Vines. NY, 1903. [viii], 451pp. Green decorated cloth. VG. **$16.00\£9.00** *(BB 12/88)*
Another copy. NY, 1903. 451pp. Partially uncut. Dec. cover. VG. **$25.00\£14.00** *(RW 6/89)*

Parkinson, J. Paradisi in Sole Paradisus Terrestris, or A Garden of All Sorts of Pleasant Flowers... Collected by John Parkinson Apothecary of London—Faithfully Reprinted from the Edition of 1629. Methuen & Co., 1904. Folio. Orig brds and canvas spine. A particularly G cpy. **$270.00\£150.00** *(DL 89/90)*

Parnell, R. The Grasses of Britain. London, 1842-1845. [xxvii, xxi], 311pp. Ry 8vo. Orig clth, ntly reprd. **$108.00\£60.00** *(WW 1/90)*

Parry, Ernest J. The Chemistry of Essential Oils and Artificial Perfumes. London, 1908. 2nd ed., rev & enlrgd. [viii], 546pp. Hd of spine trn else VG. **$45.00\£25.00** *(Hrt 12/89)*

Parsons, Frances T. According to Season. NY, 1902. [xiv], 197pp + 8pp ads. New & enlarged ed. VG. **$10.00\£6.00** *(RW 6/89)*

Parsons, M. The Wild Flowers of California. CSBD, 1921. 417pp. Spine rubbed. **$18.00\£10.00** *(GW 12/88)*

Parsons, M.E. The Wild Flowers of California, Their Names, Haunts & Habits. SF, 1897. 410pp. Crisp, clean. **$35.00\£19.00** *(PBx 11/89)*

Parsons, Samuel. Landscape Gardening Studies. NY. John Lane, 1910. 107pp. Some wear to extremities. **$40.00\£22.00** *(RW 6/89)*

Parsons, Samuel. Landscape Gardening. NY & London, 1895 (1891). 329pp. Head and tail of spine worn. **$86.50\£48.00** *(ABx 3/89)*
Another copy. NY, 1891. [xxii], 329pp. T.e.g., dec. cvr. VG. **$55.00\£31.00** *(RW 6/89)*

Parsons, Samuel. The Art of Landscape Architecture. NY, 1915. [xxi], 347pp. Pict brds, about F. **$120.00\£67.00** *(Hrt 12/89)*

Paterson, Rev. N. The Manse Garden. London, 1926. 181pp. 4 tipped-in color plates. VG. **$20.00\£11.00** *(RW 11/86)*
Another copy. London, nd (1870s?). 234pp. Decorated cloth. G. **$21.50\£12.00** *(ABx 3/89)*

Paton, J.A. Wild Flowers in Cornwall. Truro, 1968. 122pp. G. **$18.00\£10.00** *(WW 1/90)*

Pavlov, N.V. (ed). Flora of Kazakhstana. Vol IV. Alma Ata, 1961. 548pp. VG. **$30.00\£17.00** *(WH 1/90)*

Pavlov, N.V. (ed). Flora of Kazakhstana. Vol V. Alma Ata, 1961. 516pp. VG. **$30.00\£17.00** *(WH 1/90)*

Pearson, C. Arthur. Pearson's Encyclopedia of Roses. London, 1964. Rev edn. 172pp. VG/F in dj. **$9.00\£5.00** *(BLB 7/89)*

Peattie, D. A Flora of the Indiana Dunes. Field Museum, 1930. 432pp. Boards. VG. **$21.00\£12.00** *(GW 12/88)*

Peattie, Donald Culross. A Natural History of Trees of Eastern and Central North America. Boston, 1963. 606pp. VG. **$25.00\£14.00** *(WH 1/90)*

Peattie, Donald Culross. A Natural History of Western Trees. Boston, 1953. 751pp. VG. **$37.50\£21.00** *(WH 1/90)*

Peattie, Donald Culross. Green Laurels. NY, 1936. [xxiii], 368pp. Very tendr spine else VG. **$12.00\£7.00** *(Hrt 12/89)*

Peck, Charles H. Mushrooms and Their Use. Cambridge Botanical Supply Co. May, 1897. 1st bound ed. 84pp. 8vo. Sl. cover rubbing else VG. **$35.00\£19.00** *(JI 12/88)*

Peck, Charles. Annual Report of the State Botanist of NY. Albany, 1896. Large 4to. Orig green cloth, gilt. 241pp. Scarce 1st ed. Bookplate of Florence Audubon. G. **$125.00\£69.00** *(WH 1/90)*

Peck, M. A Manual of the Higher Plants of Oregon. Author, nd. 866pp. Rebound in lib buckram. **$26.00\£14.00** *(GW 8/89)*

Peckham, Ethel Anson S. (ed). Alphabetical Iris Check List. Baltimore, 1939. [vii], 582pp with frontis. Green cloth. VG. **$79.00\£44.00** *(BB 12/88)*

Peden, Rachel. Speak to the Earth. NY, 1974. 240pp. F in dj. **$4.50\£3.00** *(BLB 7/89)*

Pedersen, Ch. Christiern Pedersens Laegebog. (Malmo, 1533). Kopenhagen, 1933. Issued in facs. Clth bckd brds, laid paper, uncut. In Danish. Blck lettr. F. **$65.00\£36.00** *(PBx 11/89)*

Pedley, R. & K. Coleus. Edinburgh, 1974. 1st, 116pp. As New in dj. **$7.00\£4.00** *(BLB 7/89)*

Peet, Louis H. Trees and Shrubs of Prospect Park. NY, 1902. 237pp. 12mo. Some underlining in first prt of text. Extrems wrn, G. **$30.00\£17.00** *(CB 1/90)*

Peet, Louis Harman. Trees and Shrubs of Central Park. NY, 1903. 363pp. VG. **$45.00\£25.00** *(RW 6/89)*

Pepoon, H.S. An Annotated Flora of the Chicago Area. Chicago, 1927. 576pp. Wrpprs. G. **$36.00\£20.00** *(WW 1/90)*

Perez Arbelaez, Enrique. Plantas Medicinales y Venenosas de Columbia. Bogota, 1937. 295pp. Quarter bound in leather, green cloth is simulated lizard. Original wrapper bound in. Ex ref. library copy. Leather chafing, pages browned else VG. **$48.00\£27.00** *(BB 12/88)*

Perrault, Charles. Labyrinte de Versailles. A Paris de L'Imprimerie Royale. 1677. 1st ed. 34pp. (blank), (1) explication du plan, (1) the map, blank, 79pp, consist of the engs, each faced w/Isaac Benserade's verses, blank, (3) Table des fables, (1) colophon. Fll lthr, w/gilt lining around perimeters, orig(?) 3/4 spine rebnd on newer lthr. But for darkening 1/2" stain on fore-edge, w/minimal to no bleeding into pp very lght fox & smudges a few pp. Exceptionally clean. Very rare. **$4800.00\£2670.00** *(Hrt 12/89)*

Perring, F. The Flora of a Changing Britain. Classey, 1970. 156pp. Bent boards. **$7.75\£4.00** *(GW 12/88)*

Perring, F.H. & S.M. Walters (eds). Atlas of the British Flora. London, 1962. 432pp. 4to. Clth, rather crudely rebacked. Contents G. **$54.00\£30.00** *(WW 1/90)*

Perring, F.H., P.D. Sell and S.M. Walters. A Flora of Cambridgeshire. London, 1964. [xvi], 366pp. Clth. G. **$36.00\£20.00** *(WW 1/90)*

Perry, Amos. Water, Bog and Moisture-Loving Plants. England, ca 1930. 80pp. VG. **$11.75\£7.00** *(BLB 7/89)*

Perry, Frances. The Woman Gardener. NY, ca 1955. 384pp. Spine spotted else VG. **$9.25\£5.00** *(BLB 7/89)*

Perry, Frances. Water Gardening. London, 1947. Rev. ed. 338pp. G. **$27.00\£15.00** *(ABx 3/89)*

Perry, Mac. Landscape your Florida Home. Miami, 1972. 174pp. Orig blue/green clth. Slght edge wr; nr F. **$25.00\£14.00** *(AT 3/89)*

Perthchick, Bernard and Harriet. Flowering Trees of the Caribbean. NY, 1951. 125pp. G in dj. **$39.50\£22.00** *(ABx 3/89)*

Another copy. NY & Toronto, 1959. 3rd prntng. 4to. 125pp. G. **$16.00\£9.00** *(CB 7/89)*

Pesman, M. Walter. Meet Flora Mexicana. Globe, AZ, 1962. 278pp. Illus. cream cloth. Near fine. **$25.00\£14.00** *(BB 12/88)*

Petch, C.P. and E.L. Swann. Flora of Norfolk. London, 1968. 288pp. Sppl-ment by Swann. 1975. 94pp. Stiff wrpprs. G. **$45.00\£25.00** *(WW 1/90)*

Peters, George H. The Trees of Long Island. Farmingdale, L.I. NY, 1952. 63pp. Pub. No. 1 of the L.I. Hort. Society. Green wrappers, some fading. VG. **$14.00\£8.00** *(BB 12/88)*

Peters, Ruth Marie. Your Garden in Town. NY, 1957. 180pp. VG. **$25.00\£14.00** *(Hrt 12/89)*

Petherbridge, F.R. Fungoid and Insect Pests of the Farm. London, 1923. 2nd ed. 177pp. G. **$9.50\£6.00** *(CB 1/90)*

Pevsner, Nikolaus. The Picturesque Garden and Its Influence Outside the British Isles. Wash, 1974. 121pp. As new in dj. **$20.00\£17.00** *(Hrt 12/89)*

Pfeifer, H.W. A Taxonomic Revision of the Pentandrous Species of Aristolochia. U of Ct., 1970. 134pp. As new. **$7.00\£4.00** *(PBx 11/89)*

Pfeiffer, E. Bio-Dynamic Farming and Gardening. NY, 1943. 240pp. Cloth and contents light wear. **$10.00\£6.00** *(AB 4/89)*

Pfeiffer, Ehrenfried. Bio-dynamic Farming and Gardening. NY, 1940. 2nd ed. [ix], 240pp. Extrems wrn else VG. **$11.00\£6.00** *(CB 1/90)*

Phillips, E.P. The Genera of South African Flowering Plants. Memoir #10. 1926. Orig green clth. VG. **$72.00\£40.00** *(DL 89/90)*

Phillips, E.P. The Genera of South African Flowering Plants. Pretoria, 1951. 2nd ed. 923pp. Ry 8vo. G. **$63.00\£35.00** *(WW 1/90)*

Phillips, Henry. Pomarium Britannicum. London, 1820. 1st ed. [x] including list of subscribers, [viii], 378pp. About F in diced clf, title rebckd on spine, a bit rbbd. **$270.00\£150.00** *(Hrt 12/89)*

Phillips, Henry. Pomarium Britannicum. London, 1821. 2nd ed. 378pp. Recently rebound in marbled paper. **$171.00\£95.00** *(ABx 3/89)*

Phillips, Henry. Sylva Florifera. London, 1823. 1st ed. 2 vols. Vol I (1): [vii], 336pp. Vol II: 333pp. Orig 1/4 clth/brds, wrn. Contents sound w/minimal brwning. **$180.00\£100.00** *(Hrt 12/89)*

Phillips, Martha. The Perennial Garden. SF, 1934. 120pp. Green cloth. VG. **$12.00\£7.00** *(BB 12/88)*

Phillpotts, Eden. My Garden. London, 1906. [x], 207pp. Light fdng, light fxng else a very nice copy. Becoming scarce. **$40.00\£22.00** *(Hrt 12/89)*

Phillpotts, Eden. My Shrubs. London, 1914. [vi], 132pp. Spine faded else VG. **$40.00\£22.00** *(Hrt 12/89)*

Phillpotts, Eden. My Shrubs. London, 1915. [vi], 132pp. VG. **$75.00\£42.00** *(RW 6/89)*

Pierot, Suzanne. The Ivy Book. NY, 1974. 16mo, 164pp. As New in dj. **$7.00\£4.00** *(BLB 7/89)*

Piganiol de la Force. Description Historique de la Ville de Paris et ses Environs. Paris, 1765. 12mo. 536pp. 1/4 lthr on brds. VG. **$140.00\£78.00** *(Hrt 12/89)*

Pilat, Albert. Mushrooms. London, 1954. 340pp. Scarce. Ownr's stmp; some UK postal stmps on f.f.e.p. else VG. **$36.00\£20.00** *(Hrt 12/89)*

Pim, Sheila. The Wood and Trees. London, 1966. 256pp. G in dj. **$18.00\£10.00** *(ABx 3/89)*

Pinchot, Gifford. A Primer of Forestry. Wash., DC. USDA Div. of Forestry Bull #24. 2 vols. 1900, 1905. 88pp, 88pp. Part I: The Forest. Part II: Practical Forestry. 12mo. Green cloth. Foxing. **$21.50\£12.00** *(BB 12/88)*

Pinnery, Margaret. Miniature Rose Book. Van Nostrand, 1964. 149pp. VG in dj. **$10.00\£6.00** *(AB 4/89)*

Pipe, Ann Kimball. Bonsai. NY, 1964. 188pp. Almost as new in dj. **$20.00\£11.00** *(Hrt 12/89)*

Piper, Charles V. and R. Kent Beattie. Flora of the Northwest Coast. Lancaster, PA, 1915. 1st ed. [xiv], 420pp. Pencil notes in margins, sl cover rubbing else VG. **$75.00\£42.00** *(JI 4/89)*

Piper, Charles V. Flora of the State of Washington. Wash, 1906. 637pp. Fldg map present. Insect activity last few pp else quite G. **$36.00\£20.00** *(Hrt 12/89)*

Piper, Chas. V. Flora of the State of Washington. Wash, 1906. Orig wraps, 637pp. Scarce. VG. **$30.00\£17.00** *(WH 1/90)*

Piscataqua Papers. Piscataqua Garden Club, 1965. [viii], 148pp. Wraps. VG. **$9.00\£5.00** *(RW 6/89)*

Pizarro, C.M. Las Especies de Plantas Descritas por R.R. Philippi, en el Siglo XIX. 1960. 189pp. G. **$26.00\£14.00** *(PBx 11/89)*

Plant, Alfred. Plant's Farmers and Gardener's Almanac. St. Louis, Plant Seed Co., 1873. 64pp. Wraps and contents nearly new. **$55.00\£31.00** *(AB 4/89)*

Platt, Rutherford. This Green World. NY, 1942. 17th prntng. 219pp. VG. **$14.00\£8.00** *(Hrt 12/89)*

Plesch, Arpad. The Magnificent Botanical Library of the Stiftung fur Botanik. Vaduz, Liechtenstein, London, 1975. 3 vols. Photostats of prices laid in. Crisp and clean. (Auction). **$175.00\£97.00** *(Hrt 12/89)*

Plues, M. Rambles in Search of Wild Flowers. London, 1879. 3rd ed. [xii], 364pp. G. **$32.00\£18.00** *(WW 1/90)*

Pokorny, J. Trees of Parks and Gardens. London, 1967. 135pp. VG in repaired dj. **$10.00\£6.00** *(RW 6/89)*

Polunin, O. Flowers of the Mediterranean. Chatto, 1965. 257pp. Rubbed else VG. **$17.75\£10.00** *(GW 12/88)*

Pomerleau, R. Mushrooms of Eastern Canada & the United States. Montreal, 1951. G. **$25.00\£14.00** *(PBx 11/89)*

Pond, Barbara. A Sampler of Wayside Herbs. CT, 1974. 126pp. VG in dj. **$15.00\£8.00** *(RW 6/89)*

Pontey, W. The Profitable Planter. Huddersfield, 1808. 2nd ed. 222pp. Contemp hlf clf, slightly wrn. Cheap cpy lacking the plate. Top blank corners wrn. **$36.00\£20.00** *(WW 1/90)*

Pontey, William. The Forest Pruner. London. J. Harding, 1810. 3rd ed. [xii], 277pp. Contemp blind panelled calf front hinge split but intact. Lovely copy. **$176.50\£98.00** *(ABx 3/89)*

Pool, R.J. Flowers & Flowering Plants. NY, 1941. 2nd ed, 428pp. VG. **$10.00\£6.00** *(WH 1/90)*

Pope, Willis T. Manual of Wayside Plants of Hawaii. Tokyo, 1968. 289pp. Book VG; dj in glassine but taped to book at edges. **$22.00\£12.00** *(CB 1/90)*

Popenoe, W. Manual of Tropical and Subtropical Fruits. NY, 1920. [xv], 474pp. G. **$63.00\£35.00** *(WW 1/90)*

Popenoe, Wilson. Manual of Tropical and Subtropical Fruits. NY, 1924. 474pp. G. **$35.00\£19.00** *(WH 1/90)*

Porsild, A.E. Plant Life in the Arctic. Ottawa. Nat. Museums of Canada. March, 1951. 26pp. Wraps. VG. **$5.00\£3.00** *(JI 4/89)*

Portland, Oregon Journal. Our Garden Book. 1941. 190pp. VG. **$15.00\£8.00** *(RW 6/89)*

Post, G.E. Flora of Syria, Palestine and Sinai. Beirut, 1896. 919pp. New clth. **$81.00\£45.00** *(WW 1/90)*

Postma, C. Plant Marvels in Miniature. Day, 1961. 173pp. VG in worn dj. Folio. **$12.50\£7.00** *(GW 8/89)*

Poulsen, Svend. Poulsen on the Rose. London, 1955. 160pp. VG in lghtly worn dj. **$12.00\£7.00** *(AB 4/89)*

Powell, E.P. Hedges, Windbreaks, Shelters and Live Fences. NY, 1922. [xiii], 140pp. VG. **$35.00\£19.00** *(RW 6/89)*

Powell, Florence Lee. In the Chinese Garden. NY, 1943. 112pp. Green cloth, edges darkened else VG in yellowing, chipped dj. **$35.00\£19.00** *(CB 8/89)*

Poynter, F.N.L. A Bibliography of Gervase Markham 1568-1637. Oxford U.P., 1962. F. **$160.00\£89.00** *(PBx 11/89)*

Praeger, R.L. An Account of the Genus Sedum. London, 1967. 314pp. Wrpprs. G. **$32.00\£18.00** *(WW 1/90)*

Pratt, A. The British Grasses and Sedges. S.P.C.K., nd (c. 1864). [xviii], 136pp. Clth faded. **$36.00\£20.00** *(WW 1/90)*

Pratt, Anne. The Grasses, Sedges and Ferns of Great Britain. London, 1889. 140pp. G in original binding. **$135.00\£75.00** *(ABx 3/89)*

Pratt, Gladys Lynwall. American Garden Flowers. NY, 1943. 50pp. Clth bckd brds, scraped, edge wrn. G. **$10.00\£6.00** *(CB 1/90)*

Pratt, Merritt B. Shade and Ornamental Trees of California. Sacramento, np, 1922. 132pp. 4to. Gray cloth. Signed by author. VG. **$56.00\£31.00** *(BB 12/88)*

Preece, W.H.A. North American Rock Plants. NY, 1937. (First series). 204pp. Spine, thin portion of cvrs faded else VG. **$27.00\£15.00** *(CB 1/90)*

Preininger, Margaret. Japanese Flower Arrangement for Modern Homes. Boston, np, 1936. 4to. 38 half-tones. VG. **$10.00\£6.00** *(RW 11/86)*
Another copy. Boston, np, 1936. G. **$25.25\£14.00** *(ABx 3/89)*

Prescott, G. The Algae; A Review. Mifflin, 1968. 436pp. VG. **$20.00\£11.00** *(GW 12/88)*

Price, Rob. Johnny Appleseed Man & Myth. Bloomington, 1954. 320pp. VG in lightly worn dj. **$25.00\£14.00** *(AB 4/89)*

Price, Uvedale. On the Picturesque. Edinburgh, London, 1842. [xxvi], 586pp. Early 1/2 clf on dec brds. Ownr's plt. VG. **$220.00\£122.00** *(Hrt 12/89)*

Proske, Beatrice G. Brookgreen Gardens, Sculpture. Brookgreen, S.C. 1943 & 1955. 2 vols. Vol I: [xlviii], 510pp. Vol II: [viii], 183pp. VG. **$90.00\£50.00** *(Hrt 12/89)*

Q

Quackenbush, Alice. Perennials of Flowerland. NY, 1929. 258pp. Nearly New. **$12.00\£7.00** *(AB 2/89)*

Questel, Adrien. La Flore de la Guadeloupe. Paris, 1951. 327pp. Tall 8vo, green wrappers. Unopened. Fine. **$39.50\£22.00** *(BB 12/88)*

Quincy, J. Pharmacopeia Officinalis & Extemporanea, or A Complete English Dispensary.... London, 1739. 11th ed. 16+ 700+ 60pp. Fll clf.

Bndng fair only. Contents tight, crisp, hardly used. **$185.00\£103.00** *(PBx 11/89)*

Quinn, Vernon. Seeds: Their Place in Life and Legend. NY, 1936. [xi], 188pp. Brown cloth. VG in repaired dj. **$19.50\£11.00** *(BB 12/88)*

Another copy. NY, 1936. [xi], 188pp. VG in clean, bright dj, splitting at edges - in glassine. **$16.00\£9.00** *(CB 7/89)*

Quinn, Vernon. Shrubs in the Garden. NY, 1940. [x], 308pp. VG in worn dj. **$20.00\£11.00** *(RW 11/86)*

R

R.H.S. & A.H.S. (Joint issue). Daffodil Yearbook 1942. London. 96pp. Illus., boards. Ex-ref. lib. **$9.50\£5.00** *(BB 12/88)*

R.H.S. Journal. Vol. 39, part 1. 1912. 312pp, [xcii]. Green cloth binding worn, gold embossed spine. F.f.e.p. inscribed "James West, 1928." **$45.00\£25.00** *(BB 12/88)*

R.H.S. Journal. Vol. LIX, part 1. London, Feb. 1934. 210pp, [lxxx, xx] ads. Wraps, VG. **$24.00\£13.00** *(Hrt 12/89)*

R.H.S. The Daffodil Yearbook 1940, no. 11. London. 100pp. Gilt decorated olive cloth. VG. **$9.50\£5.00** *(BB 12/88)*

Rabenhorst's Kryptogamenflora von Deutschland. Leipzig & NY, 1930. 1963 & 1931-59. Teil 1 & 2, 2 vols. 8vo, clth. 1 is reprnt & 2 the orig ed. Scarce. G. **$90.00\£50.00** *(WW 1/90)*

Rackham, Oliver. Hayley Wood, Its History and Ecology. Cambridgeshire & Isle of Ely, 1975. [xv], 221pp. VG. **$18.00\£10.00** *(Hrt 12/89)*

Rafinesque, C. Florula Ludovicina. Hafner reprint, 1967. 178pp. VG in dj. **$30.00\£17.00** *(GW 8/89)*

Rafinesque, C.S. Forula Ludoviciana, or A Flora of the State of Louisiana. 1967. Facs of the 1817 ed. VG in dj. **$27.00\£15.00** *(DL 89/90)*

Ragon, Hurley L. Day by Day A Nature Study. Cedar Rapids, 1931. 192pp. As New in lightly worn dj. **$10.00\£6.00** *(AB 2/89)*

Ramsey, Leonidas & Charles Lawrence. The Outdoor Living Room. NY, 1932. [xiii], 126pp. VG. **$12.50\£7.00** *(RW 6/89)*

Rankin, M.M. A Girl's Garden. London, 1905. [ix], 104pp. Decorated cloth, free end paper missing else G. Small 8vo. **$10.75\£6.00** *(ABx 3/89)*

Rauh, W. Bromeliads fur Zimmer und Gewachshaus. Orig ed in 2 vols. Nice clean set in djs. **$27.00\£15.00** *(DL 89/90)*

Raven, John. A Botanist's Garden. London, 1971. 271pp. G in dj. **$32.50\£18.00** *(ABx 3/89)*

Ray, J. Flora of Cambridgeshire. London, 1975. 146pp. G. **$21.00\£12.00** *(WW 1/90)*

Ray, John. Synopsis Methodica Stirpium Britannicarum. 1724 with Flora Anglica 1754, 1759. London, 1973. Facsimiles of both these works approx 600pp. Almost As New in dj. **$45.00\£25.00** *(ABx 3/89)*

Ray, John. Travels Through the Low Countries, Germany, Italy and France. London, 1738. 2nd ed. Combined with Catalogues Stirpium: London, 1738, as it should be & A Collection of Curious Travels and Voyages: London, 1738, 2nd ed., as it should be. Hlf glt stmpd sheep, marbld brds. Some rbbng else about F for period. **$450.00\£250.00** *(Hrt 12/89)*

Rayner, M.C. Trees and Toadstools. London. Faber & Faber, 1945. 1st ed. 72pp. VG in dj. **$15.00\£8.00** *(JI 12/88)*

Read, D.H.M. The Gardener's Yearbook 1929. London, Jan. 1929. 326pp. G. **$20.00\£11.00** *(PBx 11/89)*

Record, Samuel J. Identification of the Economic Woods of the United States. NY, 1912. 1st ed. 1st 1000. [x], 146pp. VG. **$35.00\£19.00** *(JI 12/88)*
Another copy. NY, 1919. 2nd ed, rev. & enlrgd. 157pp. G. **$10.00\£6.00** *(WH 1/90)*

Redoute, P.J. The Best of Redoute's Roses. London, 1959. Clth, folio, G. **$81.00\£45.00** *(WW 1/90)*

Reed, E.N. Late Cabbage. NY, 1917. 131pp. Contents with light stains else VG. **$18.00\£10.00** *(AB 2/89)*

Reed, H. A Short History of the Plant Sciences. Chronica Botanica, 1942. 310pp. VG. **$37.50\£21.00** *(GW 8/89)*

Rehman, Elsa. The Small Place. NY, 1918. [xxii], 164pp. VG. **$15.00\£8.00** *(RW 11/86)*

Reid, D.A. Monograph of the Stipitate Stereoid Fungi. Beih. Nov. Hedw. 18, 1965. [vii], 384pp. Ry 8vo. Wrpprs, G. **$63.00\£35.00** *(WW 1/90)*

Relhan, R. Flora Cantabrigiensis. Cambridge, 1820. 3rd ed. [xi], 597pp. Brds. G. **$25.00\£45.00** *(WW 1/90)*

Repton, Humphrey. The Art of Landscape Gardening. Edited by John Nolen. Boston and NY, 1907. 252pp. This edition scarce. G. **$135.00\£75.00** *(ABx 3/89)*

Rexford, Eben. Amateur Gardencraft. Philadelphia, 1912. 300pp. As New. **$12.50\£7.00** *(AB 4/89)*

Rexford, Eben. Four Seasons in the Garden. Philadelphia, 1907. 307pp. VG in worn dj. **$12.00\£7.00** *(AB 2/89)*

Rhododendron Society Notes. London, 1916-31. Vols 1-3, vol 2 bnd in buckram, vols 1 & 3 in prts. Vol I prt 5 in xerox, else complete set. All publshd, rare. G. **$162.00\£90.00** *(WW 1/90)*

Richter, W. Orchid Care. London, 1971. 200pp. Ry 8vo. G. **$27.00\£15.00** *(WW 1/90)*

Richter, Walter. The Orchid World. Trans. and rev. by Dr. Edmund Launert. London, 1965. 292pp. G in repaired dj. **$32.50\£18.00** *(ABx 3/89)*

Ridge, Antonia. For Love of a Rose. London, 1971. 10th imp. 254pp. G in dj. **$9.00\£5.00** *(ABx 3/89)*

Ridge, Antonia. The Man Who Painted Roses. London, 1974. 1st ed. 410pp. VG in dj. **$25.00\£14.00** *(JI 7/88)*

Rinaldi, Augusto & Vassili Tyndale. The Complete Book of Mushrooms. Crown, 1974. 332pp. Tips bumpd else VG in chppd dj. **$25.00\£14.00** *(CB 1/90)*

Rion, Hanna. Let's Make A Flower Garden. NY, 1912. 208pp. VG. **$10.00\£6.00** *(AB 2/89)*

Rivers, Thomas. The Miniature Fruit Garden. NY, 1866. [x], 133pp. Top of spine chipped else VG. **$10.00\£6.00** *(RW 11/86)*

Robbins, A.R. 25 Vegetables Anyone Can Grow. NY, 1942. 219pp. As New. **$11.00\£6.00** *(AB 2/89)*

Roberts, Edith and Julia R. Lawrence. American Ferns. NY, 1941. 2nd prntng. [viii], 98pp. VG in tattered dj. **$16.00\£9.00** *(CB 1/90)*

Roberts, H. English Gardens. Collins, 1947. 48pp. VG in dj. **$11.00\£6.00** *(GW 12/88)*
Another copy. London, 1947. 48pp. VG. **$7.50\£4.00** *(RW 6/89)*

Roberts, H. Plant Hybridization before Mendel. Hafner, 1965. 374pp. VG. **$25.00\£14.00** *(GW 8/89)*

Roberts, Harry. The Book of Old-Fashioned Flowers. London, 1901. 111pp. G. **$10.75\£6.00** *(ABx 3/89)*

Roberts, Martha M. Public Gardens and Arboretums of the United States. NY, 1962. 4to. [x], 148pp. VG in dj. **$20.00\£11.00** *(RW 11/86)*

Roberts, Martha. Public Gardens and Arboretums of the U.S. NY, 1962. [x], 148pp. E.p.s. fxd, tips bmpd, head/foot of spine wrn. G. **$14.00\£8.00** *(CB 1/90)*

Robichaud, M. & B. Buell. Vegetation of New Jersey. Rutgers, 1973. 340pp. Wrs. VG. **$10.00\£6.00** *(GW 8/89)*

Robin, C. Histoire Naturelle des Vegetaux Parasites. Paris, 1853. [xvi], 702pp. 2 vols. 8vo & ry 8vo. Clth and orig brds. Rare, with an atlas which has lib stmps on verso of pltes. Some fxing of txt leaves. **$171.00\£95.00** *(WW 1/90)*

Robinson, M. The New Family Herbal. London, n.d. (c. 1870). 480pp. 12mo. Orig clth, gilt. G. **$81.00\£45.00** *(WW 1/90)*

Robinson, W. Flora and Sylva, A Monthly Review for Lovers of Garden, Woodland, Tree or Flower. 1903-05. 3 vols. 4to. Orig blue clth, well preservd. **$630.00\£350.00** *(DL 89/90)*

Robinson, W. The English Flower Garden. London, 1898. 6th ed. 832pp. G. **$45.00\£25.00** *(ABx 3/89)*

Robinson, W. The English Flower Garden. London, 1905. 9th ed. G. **$30.00\£17.00** *(WW 1/90)*

Robinson, William. Alpine Flowers for English Gardens. London, 1870. 1st ed. 392pp. Original decorated cloth, slightly worn at head and tail of spine. **$68.50\£38.00** *(ABx 3/89)*

Robinson, William. Flora and Sylva. London, April, 1903/Dec., 1905. All printed, in 3 vols. Vol. I: 324pp. Vol. II: 342pp. Vol. III: 360pp. Extremely well preserved in gilt cloth, lght fxng on plates. T.e.g. & uncut. Specific prntng of these issues, for public sale as a complete set. **$475.00\£264.00** *(Hrt 12/89)*

Robinson, William. Garden Design and Architects' Gardens. London, 1892. 73pp. Scarce. VG. **$140.00\£78.00** *(Hrt 12/89)*

Robinson, William. The English Flower Garden and Home Grounds. London, 1914. 12th ed. [xii], 989pp. VG. **$25.00\£14.00** *(RW 11/86)*

Robinson, William. The Wild Garden. London, 1894. 4th ed. 304pp. Rebound in green calf with marbled boards. A lovely copy. **$104.50\£58.00** *(ABx 3/89)*

Robyns, W. Tentamen Monographiae Vangueriae. Brussels, 1928. 359pp. Ry 8vo. Wrappers. G. **$32.00\£18.00** *(WW 1/90)*

Rock, J.F. The Ornamental Trees of Hawaii. Honolulu, 1917. 210pp. Trifle faded. G. **$72.00\£40.00** *(WW 1/90)*

Rockwell, F.F. & W.G. Breitenbucher. Gardening with Peat Moss. Boston, 1935. 72pp. VG in dj. **$5.00\£3.00** *(RW 11/86)*

Rockwell, F.F. Rock Gardens. NY, 1929. 86pp. VG. **$5.00\£3.00** *(RW 6/89)*

Rockwell, Frederick. Around the Year in the Garden. NY, 1917. [xx], 350pp. VG. **$12.50\£7.00** *(RW 6/89)*

Rodale, J.I. Complete Book of Composting. Emmaus, 1974. 1007pp. As New. **$8.00\£4.00** *(AB 2/89)*

Rodale, J.I. Encyclopedia of Organic Gardening. Emmaus, 1971. 1145pp. As New. **$6.00\£3.00** *(AB 2/89)*

Rodale, J.I. How to Grow Vegetables and Fruits the Organic Method. Emmaus, 1971. 926pp. VG in dj. **$8.00\£4.00** *(AB 2/89)*

Another copy. Emmaus, PA, 1961. 926pp. VG in dj. **$10.00\£6.00** *(RW 6/89)*

Rodale, J.I. Organic Gardening. 1st ed. 1955. Sgnd by author. F in G dj. **$16.00\£9.00** *(PBx 11/89)*

Rodway, L. The Tasmanian Flora. Hobart, 1903. [xx], 320pp. Ry 8vo. Orig brds, G. **$135.00\£75.00** *(WW 1/90)*

Roe, E.P. Success with Small Fruits. NY. Dodd, ca. 1900. 388pp. Cloth lt. wear, contents As New. **$15.00\£8.00** *(AB 4/89)*

Roger-Smith, H. Plant Hunting in Europe. Bedford, n.d. 80pp. Ex-libris plt & dealer's sm stmp, water stn dj, bck covr & last few pp. Else reasonably G. **$18.00\£10.00** *(Hrt 12/89)*

Rogers, Cameron. Trodden Glory. Santa Barbara, 1949. [ix], 130pp. VG. **$24.00\£13.00** *(BB 12/88)*

Rogers, W.S. Garden Planning. NY, 1911. [xii], 423pp. VG. **$12.50\£7.00** *(RW 11/86)*

Rogers, W.S. Planning Your Garden. NY, 1929. [xi], 301pp. VG. **$15.00\£8.00** *(Hrt 12/89)*

Rogers, W.S. Planning Your Garden. NY, 1929. [xi], 301pp. VG. **$10.00\£6.00** *(RW 6/89)*

Rogers, Walter E. Tree Flowers of Forest, Park and Street. Appleton, 1935. 4to. 500pp. Scarce. F. **$50.00\£28.00** *(WH 1/90)*

Rohde, E.S. Garden of Herbs. London, 1920. 1st ed. 232pp. G. **$32.50\£18.00** *(ABx 3/89)*
Another copy. Boston & NY, 1936. 4th US prntng. G. **$28.75\£16.00** *(ABx 3/89)*

Rohde, E.S. Garden-Craft in the Bible. London, 1927. Orig clth-bckd brds, a little fxng. G. **$45.00\£25.00** *(WW 1/90)*

Rohde, E.S. The Story of the Garden. London, 1932. Ry 8vo, G. **$36.00\£20.00** *(WW 1/90)*

Rohde, E.S. Vegetable Cultivation and Cookery. London, 1944. Faded, G. **$18.00\£10.00** *(WW 1/90)*

Rohde, Eleanour Sinclair. Garden-Craft in the Bible. London, 1927. 1st ed. 242pp. G. **$45.00\£25.00** *(ABx 3/89)*

Rohde, Eleanour Sinclair. Gardens of Delight. London, 1934. 1st ed. G. **$50.00\£28.00** *(ABx 3/89)*
Another copy. Boston, 1936. [xii], 308pp. VG. **$20.00\£11.00** *(RW 6/89)*
Another copy. Boston, 1936. 2nd, 308pp. VG in worn dj. **$20.00\£11.00** *(AB 4/89)*

Rohde, Eleanour Sinclair. Oxford's College Gardens. London, 1932. [xii], 193pp. Blue cloth, gilt spine. Bubbling to boards else VG. **$45.00\£25.00** *(CB 8/89)*

Rohde, Eleanour. Story of the Garden. Boston, 1936. 326pp. Hinges reinforced. Dj worn. **$20.00\£11.00** *(AB 4/89)*

Romanes, George John. Darwin, and After Darwin. Chicago, 1896 & 1897. 2 vols. Both 2nd eds. [xiv], 460pp, [x], 344pp. T.e.g. Very nice set bnd in orig blue clth. Ownr's sig vol I. **$45.00\£25.00** *(Hrt 12/89)*

Root, A.I. Handling Farm Produce. Medina, OH. Root, 1896. 16pp. Paper. As New. **$20.00\£11.00** *(AB 2/89)*

Roper, L. The Gardens in the Royal Park at Windsor. London, 1959. 128pp. Clth, G. **$32.00\£18.00** *(WW 1/90)*
Another copy. London, 1959. 128pp. Ownr's stmp else VG in dj. **$48.00\£27.00** *(Hrt 12/89)*
Another copy. Garden City, NY. 1959. 128pp. About F in reprd dj. **$25.00\£14.00** *(CB 1/90)*

Roper, Lanning. On Gardens and Gardening. NY, 1969. 1st, 240pp. VG in dj. **$20.00\£11.00** *(AB 2/89)*

Roper, Lanning. The Gardens in the Royal Park at Windsor. NY, 1959. 128pp. VG in dj. **$20.00\£11.00** *(RW 6/89)*

Another copy. NY, 1959. 128pp. VG in dj. **$45.00\£25.00** *(ABx 3/89)*

Rose, Louis. La Bonne Fermiere, ou Elemens Economiques. Lille, 1765. [viii], 254pp. Tree calf, raised bands, gilt. Some light damp-staining else VG. **$125.00\£69.00** *(WH 1/90)*

Rosendahl, Carl & Frederic Butters. Trees and Shrubs of Minnesota. Minneapolis, 1928. [vii], 385pp. Clipped f.f.e.p., else G in tattered dj. **$29.50\£16.00** *(BB 12/88)*

Rosenvinge, L.K. & E. Warming (eds). The Botany of Iceland. Copenhagen, 1912-49. 5 vols. in 9 (of 12) prts. Wrpprs. G. **$72.00\£40.00** *(WW 1/90)*

Ross-Craig, S. Drawings of British Plants. London, 1971. Prt 28. Ry 8vo, wrpprs, G. **$18.00\£10.00** *(WW 1/90)*

Rossi, B.V. Modern Roses in Australia. Melbourne, 1930. VG. **$45.00\£25.00** *(PBx 11/89)*

Rousseau, C. Geographie Floristique du Quebec-Labrador. Presses de L'Université Laval, 1974. 799pp. F. **$40.00\£22.00** *(PBx 11/89)*

Rousseau, J. ; M. Gauvreau; C. Morin. Bibliographie des Travaux Botaniques. Montreal, 1939. 117pp. Quite G in wrps. **$40.00\£22.00** *(Hrt 12/89)*

Rowan, Alistair. Garden Buildings. Feltham, 1968. 64pp. Wrps. VG. **$15.00\£8.00** *(Hrt 12/89)*

Rowntree, Lester. Flowering Shrubs of California and Their Value to the Gardener. Stanford, 1939. 1st ed. [xii], 317pp. Ownr's name f.f.e.p., head/foot of spine wrn. Sgnd by Rowntree on hlf-title. Scarce. **$55.00\£31.00** *(CB 1/90)*

Royal Horticultural Society. Classified List of Daffodil Names. London, 1948. 308pp. Narrow 8vo. Clth bckd brds. G. **$9.00\£5.00** *(CB 1/90)*

Royal Horticultural Society. Daffodil and Tulip Year Books. 1946-1953. 7 vols, most in dj. **$54.00\£30.00** *(DL 89/90)*

Royal Horticultural Society. Daffodil and Tulip Year Books. 1948, 49, 50. 3 vols. Orig clth. VG. **$23.00\£12.50 (DL 89/90)**

Royal Horticultural Society. Dictionary of Gardening, 1956. 2nd and last ed. 4 vols. As new in dj. **$387.00\£215.00** *(DL 89/90)*

Royal Horticultural Society. Fruit Present and Future. 1973. Vol. II. VG. **$16.00\£8.50 (DL 89/90)**

Royal Horticultural Society. Horticultural Colour Chart. British Colour Council, n.d. (1938 & 1942). 2 8vo portfolios. 200 loose sheets. F in slipcase. **$60.00\£33.00** *(BF 2/90)*

Royal Horticultural Society. The Fruit Year Book, 1947-1958. 10 vols. Complete set, VG, mostly in dj. **$81.00\£45.00** *(DL 89/90)*

Royal Society of Arts, The Story of. London, n.d. (1935). 71pp. G. **$35.00\£19.00** *(Hrt 12/89)*

Rusby, Henry H. Jungle Memories. NY, 1933. [xiii], 388pp. Ownr's sig else nr F in dec brds. Ownr's name in gilt. In dj. **$140.00\£78.00** *(Hrt 12/89)*

Russan, Ashmore & Frederick Boyle. The Orchid Seekers. London, 1893. 390pp. G. **$54.00\£30.00** *(ABx 3/89)*

Rutledge, Albert J. Anatomy of a Park. NY, 1971. [xii], 180pp. Pictorial green clth. F in chppd dj. **$20.00\£11.00** *(AT 3/89)*

Rutter, J. Here's Flowers. Golden Cockerel Press, 1937. G in dj. **$36.00\£20.00** *(WW 1/90)*

Rutter, W.P. Wheat Growing in Canada. –The United States and the Argentine. London, 1911. G. **$18.00\£10.00** *(PBx 11/89)*

Rydberg, P.A. Flora of Colorado. Fort Collins, CO. A.E.S. 1906. 1st ed. [xxii], 448pp. 8vo. Back hinge cracked else VG. **$45.00\£25.00** *(JI 4/89)*

Rydberg, Per Axel. Flora of the Prairies and Plains of Central North America. NY Bot. Grdn, 1932. 969pp. VG. **$37.50\£21.00** *(WH 1/90)*

S

Sachs, J. von. Textbook of Botany. Oxford, 1882. 2nd ed. [xii], 980pp. Hlf morocco, trifle wrn. Ry 8vo. **$54.00\£30.00** *(WW 1/90)*

Sackville-West, V. In Your Garden. London, 1951. 1st ed. 237pp. G in torn & repaired dj. **$45.00\£25.00** *(ABx 3/89)*
Another copy. London, 1951. 2nd imp., 1951. 237pp. G. **$40.00\£22.00** *(Hrt 12/89)*

Sackville-West, V. THe Garden. A Poem. London, 1946. 2nd imp. 135pp. G in dj. **$27.00\£15.00** *(ABx 3/89)*

Sackville-West, Vita. A Joy of Gardening. NY, 1958. [ix], 210pp. VG in dj. **$20.00\£11.00** *(RW 6/89)*

Sackville-West, Vita. English Country Houses. London, 1947. 47pp & index. Wrps. VG. **$20.00\£11.00** *(Hrt 12/89)*

Sackville-West, Vita. In Your Garden Again. London, 1953. 1st ed. 178pp. F in dj. **$65.00\£36.00** *(Hrt 12/89)*

Sackville-West, Vita. The Land. London, 1933. 107pp. 12mo. VG in wrn dj. **$12.00\£7.00** *(CB 1/90)*

Sackville-West, Vita. Vita Sackville-West's Garden Book. NY, 1968. 1st Am. ed. 250pp. VG in dj. **$20.00\£11.00** *(RW 11/86)*
Another copy. London, 1968. 2nd imp. 250pp. G in dj. **$18.00\£10.00** *(ABx 3/89)*

Safford, William Edwin. The Useful Plants of Guam. Washington, 1905. 416pp. USNH. Vol. IX. Presentation copy, modern blue cloth binding, text underlining & margin notes, light foxing. Else VG. **$42.00\£23.00** *(BB 12/88)*

Saito, Katsuo. Japanese Gardening Hints. Tokyo, 1969. 173pp. VG in dj. **$20.00\£11.00** *(BF 2/90)*

Salaman, R.N. The History and Social Influence of the Potato. Cambridge, 1949. [xxiv], 685pp. Spine faded, lacks f.f.e.p., jnts loose. **$54.00\£30.00** *(WW 1/90)*

Sale, Edith T. (ed). Historic Gardens of Virginia. Richmond, 1930. Rev. ed. [xxiv], 376pp. Waterstain on spine else VG in box. **$35.00\£19.00** *(RW 6/89)*

Salinger, Margaretta. Flowers. The Flower Piece in European Painting. NY, 1949. 1st Am. ed. 27pp. VG in dj. **$25.00\£14.00** *(RW 11/86)*

Salisbury, E.J. Downs and Dunes. London, 1952. [xiii], 328pp. G. **$45.00\£25.00** *(WW 1/90)*

Salisbury, E.J. The Living Garden. London, 1935. 2nd prntng. 338pp. Ownr's name f.f.e.p., some fxng. Lght scrtchng to covrs. **$15.00\£8.00** *(CB 1/90)*

Salisbury, E.J. The Living Garden. London, 1936. 4th imp. 338pp. G in dj. **$32.50\£18.00** *(ABx 3/89)*

Salmon, C.E. Flora of Surrey. London, 1931. 688pp. G. **$81.00\£45.00** *(WW 1/90)*

Sampson, H.C. The Coconut Palm. London, 1923. [xiv], 262pp. Tall 8vo. Brick colored cloth, illus. & gilt embossed. Lower corners bumped, light foxing else VG. Signed presentation copy. **$95.00\£53.00** *(BB 12/88)*

Samuel, M. & Casey, H. Wild Flower Hunter: The Story of Ellis Rowan. Australia, 1961. 151pp. VG. **$20.00\£11.00** *(GW 8/89)*

Sanders, T.W. The Amateur's Greenhouse. London, nd (1910). 448pp. Dec brds, tender. **$18.00\£10.00** *(Hrt 12/89)*

Sanders, T.W. The Encyclopedia of Gardening. London, n.d. (1919). [xv], 482pp. Brwn pict clth. VG. **$11.00\£6.00** *(CB 1/90)*

Sanderson, Dwight. Insect Pests. NY, 1912. 684pp. Light wear, contents VG. **$18.00\£10.00** *(AB 2/89)*

Sandford, E. A Manual of Exotic Ferns and Selaginella. London, 1894. 286pp. Trifle wrn, bndng spotted. **$27.00\£15.00** *(WW 1/90)*

Sanger, M.B. Billy Bartram. NY, 1972. [xii], 207pp. As new in dj. **$35.00\£19.00** *(Hrt 12/89)*

Santapau, H. The Flora of Purandhar. India, 1957. 158pp. VG. **$16.50\£9.00** *(GW 8/89)*

Sargent, C. The Scientific Papers of Asa Gray 1834-1867. 503pp. Kraus reprint. 1968. VG. **$25.00\£14.00** *(GW 12/88)*

Satriana, M.J. Large Scale Composting. NJ, 1974. 269pp. Cloth and contents VG. **$15.00\£8.00** *(AB 4/89)*

Sattler, R. Organogenesis of Flowers: A Photographic Text-Atlas. Toronto, 1973. 207pp. VG in worn dj. **$25.00\£14.00** *(GW 8/89)*

Saunders, Charles F. Western Flower Guide. NY, 1922 (1917). 286pp. Oblong small 12mo in worn slipcase. Decorated black flexible imitation leather covers. **$17.00\£9.00** *(BB 12/88)*

Saunders, Charles Francis. The Story of Carmelita. Pasadena. 1928, 54pp. 12mo. Green decorated cloth. Inscr. on title page, else VG. **$19.00\£11.00** *(BB 12/88)*

Saunders, Charles. Trees and Shrubs of California Gardens. NY, 1926. 323pp. VG. **$15.00\£8.00** *(AB 2/89)*

Savidge, J.P., V.H. Heywood and V. Gordon. Travia's Flora of South Lancashire. Liverpool, 1963. G. **$27.00\£15.00** *(WW 1/90)*

Savonius, Moira. Mushrooms and Fungi. NY, 1973. 4to, 72pp. F in dj. **$7.50\£4.00** *(WH 1/90)*

Scagel, R. An Annotated List of the Marine Algae of British Columbia and Northern Washington. Canada, 1957. 289pp. Wrs. **$17.50\£10.00** *(GW 8/89)*

Schabol, l'Abbe Roger. La Pratique du Jardinage. Paris, 1774. Nouvelle Edition. 12mo. 2 vols. Vol I: 504pp. Vol II: 5 fldg plts. Wrn contemp brds, else very clean set. **$320.00\£178.00** *(Hrt 12/89)*

Schelpe, E.A. An Introduction to the South African Orchids. Cape Town, 1966. 112pp. Demy 4to. G. **$36.00\£20.00** *(WW 1/90)*

Schling, Max. Everyman's Garden. NY, 1935. 1st ed. [xii], 106pp. Orig green clth, gilt titles. Some sunning to spine edges else very nice cpy. **$18.00\£10.00** *(AT 3/89)*

Schmidt, Hubert. Agriculture in New Jersey. NJ, 1973. 335pp. VG in dj. **$30.00\£17.00** *(AB 4/89)*

Schneider, G. Choice Ferns for Amateurs. London, 1905. 330pp. Trifle used. **$30.00\£17.00** *(WW 1/90)*

(Schola Salernitana). Schola Salernitana, Sive, de Conservanda Valetudine. . . Arnoldi Villanovani. . . Nova Editio, Augustae Vindelicorum (Augsburg), 1753. 558pp. Full clf, gilt spine, 2-col title. VG. **$235.00\£131.00** *(PBx 11/89)*

(Schola Salernitana). De Conservanda Bona Valetudine Opusculum Schola Salernitanae. Franc(furt), Egenolph, 1554. Title in red & black. Rubricated frontis. Many wood cuts. Early vellum, 177pp. Shows age but little wear. **$1,500.00\£833.00** *(PBx 11/89)*

Schomburgk, R. Catalogue of the Plants under Cultivation in the Government Botanic Garden. Adelaide, 1871. [xiii], 200pp. Brds. Hole through upper margin. **$36.00\£20.00** *(WW 1/90)*

Schott, H. Genera Filicum. Vienna, 1834. 31 leaves of text with 14 plts. 4to, new brds. Some fxing. **$63.00\£35.00** *(WW 1/90)*

Schroeter, C. Flora des Sudens. Zurich, 1936. G. **$36.00\£20.00** *(WW 1/90)*

Schuler, E. Gardens around the World. Abrams, 1964. 256pp. VG in dj. **$20.00\£11.00** *(GW 8/89)*

Schuler, Stanley. The Winter Garden. NY, 1972. 244pp. Ink # f.f.e.p. else about F. **$18.00\£10.00** *(Hrt 12/89)*

Schultes, Richard E. & Arthur S. Pease. Generic Names of Orchids. NY. Academic Press, 1963. [xiii], [2], 331pp. VG in dj. **$34.00\£19.00** *(BB 12/88)*

Schulz, Ellen D. Texas Wild Flowers. Chicago, 1928. [xi], 505pp. Decorated green cloth. Fine. **$33.00\£18.00** *(BB 12/88)*

Schulz, Peggie (ed). Gesneriads and How to Grow Them. Grandview, Missouri, 1967. 1st ed. 352pp. VG in dj. **$15.00\£8.00** *(JI 7/88)*

Schuster/Steere/Thomson. The Terrestrial Cryptogams of Northern Ellesmere Island. N.M. Canada Bull. 164. 1959. **$11.00\£6.00** *(PBx 11/89)*

Schweinfurth, C. Orchids of Peru. 4 vols. Fieldiana, 1961. 1005pp. 194 figs. Wrs. **$65.00\£36.00** *(GW 8/89)*

Schweinfurth, G. Abyssinische Pflanzennamen. Berlin, 1893. 84pp, 4to. Marbled boards. VG. **$28.00\£16.00** *(BB 12/88)*

Scoggan, H.J. Flora of Manitoba. Ottawa, 1957. 619pp, stiff wrappers. F. **$27.50\£15.00** *(WH 1/90)*

Scopoli, J.A. Flora Carniolica. Vienna, 1760. [xxii], 607pp. New clf. Blnd stmp on title-pg, some slt fxing. **$450.00\£250.00** *(WW 1/90)*

Scott, D.H. Extinct Plants and Problems of Evolution. London, 1924. [xiv], 240pp. London, 1924. G. **$27.00\£15.00** *(WW 1/90)*

Scott, Frank J. The Art of Beautifying Suburban Home Grounds of Small Extent. NY, 1881. 618pp. Orig bndng tight, but scuff & sm stains. Except for light dmp stning bottom/side edges interior. VG. **$220.00\£122.00** *(Hrt 12/89)*

Scott, Frank J. The Art of Beautifying Suburban Home Grounds. NY, 1881. 618pp. Gilt dec green cloth, t.e.g. Top of spine wrn away, frnt jnt cracked. **$85.00\£47.00** *(MD 2/90)*

Scott, M.H. & A. Edgar Beresford. Houses and Gardens. London, 1933. 4to. 294pp. **$95.00\£53.00** *(RW 11/86)*

Scott, Owen R. St. Clair Parkway. Np, 1967. 67pp. Ownr's sig else VG. **$25.00\£14.00** *(Hrt 12/89)*

Scott, S.H. The Observer's Book of Cacti. London & NY, 1964. 6th prntng. 12mo. 159pp. VG in G dj. **$7.00\£4.00** *(CB 8/89)*

Sealy, J. Robert. A Revision of the Genus Camellia. London, 1958. 239pp. G. **$126.00\£70.00** *(ABx 3/89)*

Sedgwick, Mabel Cabot. The Garden Month by Month. Garden City, 1907. [xvii], 516pp. VG. **$17.00\£9.00** *(CB 1/90)*

Sedgwick, Mabel. The Garden Month by Month. Garden City, 1907. 516pp. Cloth and contents VG. **$15.00\£8.00** *(AB 4/89)*

Another copy. NY, 1907. [xviii], 516pp. VG. **$25.00\£14.00** *(RW 6/89)*

Setchell, W.A. & N.L. Gardner. Algae of Northwestern America. CA, 1903. 397pp. Wraps, VG. **$20.00\£11.00** *(GW 12/88)*

Setchell, W.A. & N.L. Gardner. The Marine Algae of the Pacific Coast of N. America. CA, 1919. 3 vols. 739pp. Wraps, VG. **$50.00\£28.00** *(GW 12/88)*

Seward, A.C. Fossil Plants. Cambridge, 1898-1919. 4 vols, vol. 2 spine faded. **$135.00\£75.00** *(WW 1/90)*

Seward, A.C. Plant Life through the Ages. Cambridge, 1931. [xxi], 601pp. Ry 8vo. Clth. G. **$54.00\£30.00** *(WW 1/90)*

Seward, Barbara. The Symbolic Rose. NY, 1960. [viii], 233pp. G. **$35.00\£19.00** *(Hrt 12/89)*

Sewell, Cornelius V.V. Common Sense Gardens. NY, 1906. [xviii], 396pp. Dec. cvr. VG. **$20.00\£11.00** *(RW 6/89)*

Seymour, E.L.D. The Home Garden. NY, 1917. 64pp. VG. **$15.00\£8.00** *(RW 6/89)*

Seymour, E.L.D. Wise Garden Encyclopedia. NY, 1956. 1380pp. Cloth and contents Nearly New. **$10.00\£6.00** *(AB 4/89)*

Shafer, Sara A. A White-Paper Garden. Chicago, 1910. 292pp. VG. **$20.00\£11.00** *(AB 4/89)*

Another copy. Chicago, 1910. [xxi], 292pp. VG. **$10.00\£6.00** *(RW 11/86)*

Another copy. Chicago, 1910. 292pp. Very light foxing else G. **$39.50\£22.00** *(ABx 3/89)*

Shannon, Fred. Farmer's Last Frontier, Agriculture 1860-1897. NY, 1945. 434pp. VG in worn dj. **$12.00\£7.00** *(AB 4/89)*

Sharples, Ada White. Alaska Wild Flowers. Stanford U Press, 1973. [viii], 156pp. VG in dj. **$15.00\£8.00** *(JI 4/89)*

Shaw, George R. The Pines of Mexico. Jamaica Plain, 1961 (1909). [4], 29pp, [1]. 4to. Rust colored wrappers. As New. **$35.00\£19.00** *(BB 12/88)*

Shearn, W.B. et al. The Practical Fruiterer & Florist. London, 1935. 3 vols. G. **$60.00\£33.00** *(PBx 11/89)*

Shelton, L. Continuous Bloom in America. Scribner, 1915. 145pp. VG. **$15.00\£8.00** *(GW 12/88)*

Shelton, Louise. Beautiful Gardens in America. NY, 1916. 2nd ed. [xvi], 341pp. Tendr bck brd else almost F. **$70.00\£39.00** *(Hrt 12/89)*

Shepherd, Roy E. History of the Rose. NY, 1954. 264pp. VG in chipped dj. **$25.00\£14.00** *(AB 4/89)*

Another copy. 264pp. Coleman reprnt. 1954. VG in dj. **$20.00\£11.00** *(GW 12/88)*

Sheppard, J.C. & G.A. Jellicoe. Italian Gardens of the Renaissance. London, 1953. 207 b/w illus, [viii], 25pp. Ownr's sig, dmpness/slt bowing of covrs else quite G. **$55.00\£31.00** *(Hrt 12/89)*

Sheppard, R.W. Enjoying Nature in a Garden. Niagara Falls, Ont., 1975. 111pp. Wraps. F. **$12.00\£7.00** *(Hrt 12/89)*

Sherlock, Chesla C. City and Suburban Gardening. NY, 1928. [x], 186pp. VG. **$15.00\£8.00** *(Hrt 12/89)*

Shetler, S. The Komarov Botanical Institute. 250 Years of Russian Research. Smithsonian Public. 4687. 1967. G in dj. **$18.00\£10.00** *(PBx 11/89)*

Shetler, S. The Komarov Botanical Institute: 250 Years of Russian Research. Smith, 1967. 240pp. **$7.50\£4.00** *(GW 12/88)*

Shewell-Cooper, W.E. The Royal Gardeners. London, 1952. [xiii], 66pp. Ownr's inscrp f.f.e.p. else G in chppd dj. **$13.00\£7.00** *(CB 1/90)*

Shewell-Cooper, W.E. The Royal Gardeners. London, 1952. Sq 8vo. [xii], 66pp. VG in worn dj. **$10.00\£6.00** *(RW 6/89)*

Shreve, F. A Montane Rain-Forest: A Contribution to the Physiological Plant Geography of Jamaica. Carnegie, 1914. 110pp. A few library stamps. **$40.00\£22.00** *(GW 8/89)*

Shreve, F. The Cactus & Its Home. Williams, 1931. 195pp. VG in worn dj. Signed. **$20.00\£11.00** *(GW 12/88)*
Another copy. Balto, 1931. 195pp. F. **$20.00\£11.00** *(WH 1/90)*

Sim, T.R. Tree Planting in South Africa. Pietermaritzburg, 1927. 452pp. G. **$54.00\£30.00** *(WW 1/90)*

Simmonds, A. A Horticultural Who Was Who. London. R.H.S., 1948. 80pp. G in dj. **$16.25\£9.00** *(ABx 3/89)*

Simmonds, P.L. The Commercial Products of the Vegetable Kingdom. London, 1854. 668pp. Hlf clf on marbld brds, mrbld edges. VG. **$110.00\£61.00** *(Hrt 12/89)*

Simmons, Adelma G. Herb Gardening in Five Seasons. Princeton, 1964. [xx], 353pp. VG in dj. **$15.00\£8.00** *(RW 6/89)*

Simmons, H.C. A Survey of the Phytogeography of the Arctic American Archipelago. Lund, 1913. 183pp. Brds. G. **$45.00\£25.00** *(WW 1/90)*

Simmons, James R. The Historic Trees of Massachusetts. Boston, 1919. [xxi], 139pp. VG. **$20.00\£11.00** *(RW 6/89)*

Simmons, James Raymond. The Historic Trees of Massachusetts. Boston, 1919. 139pp. About F, 1/4 lthr on brds. **$27.00\£15.00** *(Hrt 12/89)*
Another copy. Boston, 1919. [xxi], 139pp. 2nd prntng. Clth bckd brds. VG. **$20.00\£11.00** *(CB 1/90)*

Simon, L. & P. Cochet. Nomenclature de Tous les Noms de Roses. Paris, 1906. Tall 8vo. Pbnd. 177pp. G. **$35.00\£19.00** *(PBx 11/89)*

Simonds, John Ormsbee. Landscape Architecture. NY, 1961. [xi], 244pp. VG in dj. **$20.00\£11.00** *(RW 6/89)*
Another copy. NY, 1961. 244pp. Grey cloth, slightly wrn dj else VG. **$15.00\£8.00** *(MD 2/90)*

Simonds, John Ormsbee. Landscape Architecture. NY, 1961. 244pp. VG in wrn dj. **$55.00\£31.00** *(Hrt 12/89)*

Simonsen, R. (ed). First Symposium on Recent and Fossil Marine Diatoms. Bremerhaven, 1970. [viii], 296pp. Ry 8vo. Wrpprs. G. **$45.00\£25.00** *(WW 1/90)*

Simpfendorfer, K.J. An Introduction to Trees for South Eastern Australia. Melbourne, 1975. 1st ed. [xiv], 377pp. 4to. Green cloth. VG. **$57.50\£32.00** *(BB 12/88)*

Simpson, Charles T. Ornamental Gardening in Florida. Little River, FL, 1926. [x], 243pp. VG, some light foxing. **$23.00\£13.00** *(BB 12/88)*

Simpson, Lesley Byrd (trans & ed). Journal of Jose Longinos Martinez. SF, 1961. Ltd to 1000 copies. [xviii], 118pp. Unopened. VG. **$65.00\£36.00** *(JI 12/88)*

Simpson, N.D. A Bibliographical Index of the British Flora; Including Floras, Herbals, Periodicals, Societies. . . 1960. Priv prntd. Ltd to 750 copies. 429pp. VG. **$60.00\£33.00** *(PBx 11/89)*

Singer, R. Die Rohrlinge. 2 vols. I: Die Boletaceae. II: Die Boletoideae und Atrobilomycetaceae. 1965-67 and 2 folders with col and other plts. (Die Pilze Mitteleuropas Series). Well preservd set. **$108.00\£60.00** *(DL 89/90)*

Singer, R. The Genera Marasmiellus, Crepidotus and Simocybe in the Neotropics. Beih. Nov. Hedw. 44, 1973. [iv], 518pp. Ry 8vo. Wrpprs. G. **$63.00\£35.00** *(WW 1/90)*

Singleton, Esther. The Shakespeare Garden. NY, 1931. New ed. 366pp. G. **$48.50\£27.00** *(ABx 3/89)*

Sitterson, J.C. Sugar Country: The Cane Sugar Industry in the South, 1753-1950. U of Ky Press, 1953. 414pp. G. **$20.00\£11.00** *(PBx 11/89)*

Sitwell, S. & W. Blunt. Great Flower Books, 1700-1900. 1956. Royal folio, orig 1/2 clth. G in slightly trn dj. Scarce. **$810.00\£450.00** *(DL 89/90)*

Sitwell, Sacheverell. Old Fashioned Flowers. London, 1939. 193pp. Sm owner's label else VG in aging dj. **$75.00\£42.00** *(Hrt 12/89)*

Sitwell, Sacheverell. Old Fashioned Flowers. London, 1939. 1st ed. 193pp. 4to. Light shelf wear, lower front corner bumped. **$39.00\£22.00** *(BB 12/88)*

Another copy. London, 1948. 2nd ed. 4to. 193pp. VG in dj. **$25.00\£14.00** *(RW 11/86)*

Another copy. London, 1948. Reprint (1939). G in dj. **$45.00\£25.00** *(ABx 3/89)*

Skinner, Charles M. Little Gardens. NY, 1904. 1st ed. [ix], 250pp. Decorative cloth. VG. **$20.00\£11.00** *(RW 11/86)*

Slate, George. Lilies for American Gardens. NY, 1939. 1st ed. 258pp. Covers slightly marked else G in dj. **$41.00\£23.00** *(ABx 3/89)*

Sleumer, H. An Account of Rhododendron in Malesia. Groningen, The Netherlands. Nd, partly a reprint from Flora Malesiana, Jan. 1966. [iv], 473-674pp. Wraps. 8vo. Sl grubby cover else nice copy. **$35.00\£19.00** *(JI 12/88)*

Slingerland, M. & C. Crosby. Manual of Fruit Insects. NY, 1915. 503pp. Cloth light wear, inner hinges repaired else VG. **$18.00\£10.00** *(AB 4/89)*

Another copy. NY, 1924. [xvi], 503pp. Dec. cvr. VG in dj. **$15.00\£8.00** *(RW 6/89)*

Slosson, Elvenia (ed). Pioneer American Gardening. NY, 1951. [xiv], 306pp. VG in dj. **$15.00\£8.00** *(RW 6/89)*

Smiles, Samuel. Robert Dick Baker of Thurso. London, 1878. 1st ed. [xxii], 444pp. Unopened. F. **$50.00\£28.00** *(JI 7/88)*

Smiley, Nixon. Subtropical Gardening in Florida. (Coral Gables), U of Miami Press, 1951. [x], 182pp. About VG. **$20.00\£11.00** *(AT 3/89)*

Smiley, Nixon. Tropical Planting and Gardening for South Florida and the West Indies. Coral Gables: U of Miami, 1960. 1st ed. [viii], 202pp. Orig pict clth, sgnd by author. Slght cvr wr, else VG. **$25.00\£14.00** *(AT 3/89)*

Smith, Alexander H. & Harry D. Thiers. The Boletes of Michigan. Ann Arbor, 1971. 1st ed. [iv], 436pp. VG in dj. **$30.00\£17.00** *(JI 12/88)*

Smith, Alexander. The Mushroom Hunter's Field Guide. U of Mich, 1971. Rev and enlrgd. Tips wrn else VG. **$9.00\£5.00** *(CB 1/90)*

Smith, Alice R. Huger & Herbert Ravenel Sass. A Carolina Rice Plantation of the Fifties. NY, 1936. [xii], 97pp. About F. **$55.00\£31.00** *(Hrt 12/89)*

Smith, E. Bacteria in Relation to Plant Disease. 2 vols. 570pp. Carnegie, 1911. Folio. Worn wrs. Scarce. **$45.00\£25.00** *(GW 12/88)*

Smith, Emory E. The Golden Poppy. Palo Alto, 1902. 231pp. Green cloth, gold stamping. Book plate of Jerome A. Hart. VG. **$56.00\£31.00** *(BB 12/88)*

Smith, G. Cryptogamic Botany. Vol. II. Bryophytes & Pteridophytes. McGraw, 1938. 380pp. Bookplate removed else VG. **$20.00\£11.00** *(GW 12/88)*

Another copy. Vol. I. NY, 1938. 545pp. VG. **$17.50\£10.00** *(WH 1/90)*

Smith, G.M. The Freshwater Algae of the United States. NY, 1950. 2nd ed. 719pp. G. **$45.00\£25.00** *(WW 1/90)*

Smith, H. Clifford. Sulgrave Manor and the Washingtons. NY, 1933. 260pp. Lght fxng prelims else VG in dj. **$60.00\£33.00** *(Hrt 12/89)*

Smith, H. Hamel & F.A.G. Pape. Coco-Nuts: The Consols of the East. London, nd (ca. 1914). [lxviii], 644pp. Wear at head and tail of spine. **$28.00\£16.00** *(BB 12/88)*

Smith, Herbert H. Brazil, the Amazons and the Coast. NY, 1879. 1 fldg map, [xv], 644pp. (4)pp advts. Sl tender in gilt dec brds else VG. **$110.00\£61.00** *(Hrt 12/89)*

Smith, J. Ferns. London, 1896. 2nd ed. [xv], 450pp. Cr 8vo. Trifle used. **$27.00\£15.00** *(WW 1/90)*

Smith, J. Fruits & Farinacea, The Proper Food of Man. NY, 1856. 314pp. Mdrn bndng. Crisp. **$120.00\£67.00** *(PBx 11/89)*

Smith, L. Bermuda's Oldest Inhabitants: Tales of Plant Life. Salmon, 1963. 80pp. VG. **$15.00\£8.00** *(GW 8/89)*

Another copy. Sevenoaks, KY, 1950. 3rd ed. 80pp. Sm 4to, decorated cloth. VG in torn dj. **$32.50\£18.00** *(BB 12/88)*

Smith, Lyman B. The Bromeliads. NY, 1969. 1st ed. 56pp. Nice copy in rubbed dj with several small tears. **$35.00\£19.00** *(JI 12/88)*

Smith, R. & L. Downs. Flora Neotropica: Pitcairnioideae. Hafner, 1974. 658pp. Scuffed else VG. **$25.00\£14.00** *(GW 8/89)*

Smith, R.E.C. and E. Ledbrook. A Garland of Shakespeare's Flowers. London, 1911. [xi], 104pp. G. **$36.00\£20.00** *(WW 1/90)*

Smith, W.G. Synopsis of the British Basidiomycetes. British Museum, 1908. 531pp. G. **$18.00\£10.00** *(WW 1/90)*

Smithers, L.A. Lodgepole Pine in Alberta. Ottawa, 1961. 156pp. Wraps. VG. **$15.00\£8.00** *(JI 12/88)*

Smithsonian Scientific Series. Old and New Plant Lore, A Symposium. Vol. II. Np, 1931. (16), 390pp. Dec brds. F. **$40.00\£22.00** *(Hrt 12/89)*

Smythe, F.S. The Valley of Flowers. London, 1947. 2nd ed. 318pp. Clth, trifle faded. G. **$36.00\£20.00** *(WW 1/90)*

Snell, W.H. & E.A. Dick. The Boleti of Northeastern North America. London, 1970. 320pp. Clean, 2nd hnd. **$54.00\£30.00** *(WW 1/90)*

Solly, V.N. Gardens for Town and Suburb. NY, 1926. 1st U.S. 112pp. Sm stain on clth/brds else VG. **$35.00\£19.00** *(Hrt 12/89)*

Somerville, W. How a Tree Grows. Oxford UP, 1927. 212pp. G. **$8.00\£4.00** *(PBx 11/89)*

Sorensen, C. The Origins of Garden Art. Copenhagen, 1963. The Danish Architectural Press. Ltd. ed. 485/500. 65pp. Ownr's, a recipient, stmp else F. **$75.00\£42.00** *(Hrt 12/89)*

Sowerby, Arthur de Carle. Nature in Chinese Art. NY, 1940. 203pp. Some rbbng else VG. **$70.00\£39.00** *(Hrt 12/89)*

Sowerby, J. & J.E. Smith. English Botany: Bryophyta. London, 1790-1814. 4 vols. Hlf clf. 1 vol rebckd. Wanting plts & text 3pp, text 2240 in 4th vol. 15 plts w/old lib stmps, some faint water staining. 1st 3 vols G but 8 plts w/old lib stmps, 1 stned mrgn, 2 plts lack text. **$630.00\£350.00** *(WW 1/90)*

Sowerby, J. & J.E. Smith. English Botany: Filices. London, 1790-1814. Hlf clf. 54 hand-colored plts with descrp text. Wanting 1 plt & text. Else complete. Old lib stmps on reverse of a few plts. G. **$117.00\£65.00** *(WW 1/90)*

Sowerby, J. and J.E. Smith. Supplement to English Botany, Vol. 2. London, 1831-35. 8vo, uncut and unopened. Orig wrpprs as issued. G. **$63.00\£35.00** *(WW 1/90)*

Sowerby, J. and J.E. Smith. Supplement to English Botany, Vol. I. London, 1829-31. 8vo, uncut and unopened. In orig wrpprs as issued. G. **$63.00\£35.00** *(WW 1/90)*

Sowerby, J., J.E. Sowerby et al. English Botany or Coloured Figures of British Plants. Roxburgh, 1902. Special reissue. 3rd ed. Complete in 12 vols. T.e.g., maroon mor., tight and clean set. **$1,170.00\£650.00** *(DL 89/90)*

Sowerby, J.E. British Wild Flowers. 1894. Red polishd hlf clf on marbld brds. A.e.g. hand-colored frontis, 89 hand-colored plts. Crisp, F. **$275.00\£153.00** *(PBx 11/89)*

Sowerby, John E. The Ferns of Great Britain. London, 1855. 87pp. 1/4 lthr/brds, frnt covr loose, f.f.e.p. missing. Slght offset & fxng several plts else quite G. **$147.00\£82.00** *(Hrt 12/89)*

Speechly, W. A Treatise on the Culture of the Pine Apple and the Management of the Hot-house. York, 1779. [xvii, v], 186pp. Ry 8vo. 1st ed. Some light soiling; libr bkplte no stmps. Rare. **$324.00\£180.00** *(WW 1/90)*

Spencer, Gulford L. A Handbook for Cane-Sugar Manufacturers and Their Chemists. NY, 1920. 7th, rev, rewritten & enlrgd. [xix], 560pp. Sl wear; sft covrs. VG. **$20.00\£11.00** *(Hrt 12/89)*

Spender, R.E.S. & L.F. Pesel. Iris Culture for Amateaurs. London, 01937. G. **$21.00\£12.00** *(WW 1/90)*

Spiers, Rdv. William. Rambles and Reveries of a Naturalist. London, 1890. 1st ed. 256pp. 8vo. VG. **$65.00\£36.00** *(JI 7/88)*

Spillman, W.J. Application of Some of the Principles of Heredity to Plant Breeding. Wash. GPO, 1911. 76pp. Wraps & contents As New. **$18.00\£10.00** *(AB 4/89)*

Spry, Constance. Art of Arranging Flowers. NY, 1950. 69pp. VG in dj. **$12.00\£7.00** *(AB 2/89)*

Spry, Constance. Favourite Flowers. 1959. 1st ed. Nice cpy in dj. **$21.00\£11.50 (DL 89/90)**

Spry, Constance. Flower Decoration. NY, 1938. 152pp. Cloth and contents a bit stained and worn. Signed. **$9.00\£5.00** *(AB 4/89)*

Another copy. London, nd. 152pp. VG. **$15.00\£8.00** *(AB 2/89)*

Spry, Constance. Garden Notebook. 1940. 1st ed. VG. Scarce. **$19.00\£10.50 (DL 89/90)**

Spry, Constance. Garden Notebook. NY, 1940. 1st ed. [xi], 217pp, [ix]. G. **$18.00\£10.00** *(ABx 3/89)*

Another copy. NY, 1940. 1st ed. [xii], 217pp, [ix]. VG in dj. **$20.00\£11.00** *(RW 6/89)*

Spry, Constance. Party Flowers. London, 1955. 1st. 80pp. VG in dj. **$15.00\£8.00** *(AB 2/89)*

Spry, Constance. Simple Flowers—A Millionaire for a Few Pence. 1st ed. 1957. VG in dj. **$5.00\£3.00** *(DL 89/90)*

Spry, Constance. Summer and Autumn Flowers. London, 1951. 1st ed. 116pp. G. **$18.00\£10.00** *(ABx 3/89)*

Spry, Constance. Winter and Spring Flowers Arranged by Constance Spry. London, 1951. 1st ed. 140pp. Light mark on back cover. **$18.00\£10.00** *(ABx 3/89)*

Spry, Constance. Winter and Spring Flowers. 1951. 1st ed. VG in dj. **$16.00\£8.50 (DL 89/90)**

St. Barbe Baker, Richard. Green Glory. London, 1948. 239pp. As new in older dj. **$30.00\£17.00** *(Hrt 12/89)*

St. John, H. Flora of SE Washington & of Adjacent Idaho. SBC, 1937. 531pp. Ink notes on flyleaf else VG. **$18.00\£10.00** *(GW 12/88)*

Stamp, D. Man and the Land. New Naturalist, 1955. [xvi], 272pp. 1st ed. Slightly wrn dj. Inscrip on e.p. **$45.00\£25.00** *(WW 1/90)*

Standley, P. Flora of Glacier National Park, Montana. USNH, 1921. 438pp. Wrs. Some chipping. **$16.00\£9.00** *(GW 12/88)*

Standley, P.C. Flora of the Lancetilla Valley, Honduras. Chicago, 1931. 418pp. Orig wrpprs, G. **$32.00\£18.00** *(WW 1/90)*

Stanford, K.C. A Garden of South African Flowers. Cape Town, nd. 63pp. Green cloth. VG in dj. **$22.00\£12.00** *(BB 12/88)*

Stebbing, M.E. Colour in the Garden. London, 1947. Rev. ed. [x], 117pp. 4to. G. **$21.50\£12.00** *(ABx 3/89)*

Steele, Fletcher. Design in the Little Garden. Boston, 1924. [ix], 124pp. VG in dj. **$10.00\£6.00** *(RW 11/86)*

Steele, Fletcher. Gardens and People. Boston, 1964. 1st prntng. [ix], 221pp. Sgnd by Steele. F in scffd dj. **$22.00\£12.00** *(CB 1/90)*

Steffek, Edwin. Wild Flowers and How to Grow Them. NY, 1973. 192pp. VG in wrn dj. **$10.00\£6.00** *(CB 1/90)*

Steffek, Edwin. Wild Flowers and How to Grow Them. NY, 1973. 192pp. F in dj. **$7.50\£4.00** *(WH 1/90)*

Stemler, Dorothy. Roses of Yesterday and Today. CA, 1962. 76pp. Wraps & contents VG. **$11.00\£6.00** *(AB 4/89)*

Stemler, Dorothy. Roses of Yesterday and Today. CA, 1966. 72pp. Wraps with light wear, else As New. With overlay. **$10.00\£6.00** *(AB 4/89)*

Stemler, Dorothy. Roses of Yesterday and Today. CA, 1968. 72pp. Wraps As New with overlay. **$9.00\£5.00** *(AB 4/89)*

Step, E. & W. Watson. Favourite Flowers of Garden and Greenhouse. 1896-97. 4 vols. Scarce. Well preservd set in orig clth. **$810.00\£450.00** *(DL 89/90)*

Step, E. Toadstools and Mushrooms of the Countryside. London, 1913. [xvi], 143pp. Few pencil annotations. G. **$21.00\£12.00** *(WW 1/90)*

Step, Edward. Wayside and Woodland Blossoms. London, nd. 12mo. [xiii], 176pp. VG. **$15.00\£8.00** *(RW 6/89)*

Step, Edward. Wild Flowers Month by Month in Their Natural Haunts. London, 1905. 1st ed. Vol I: 200pp. Vol II: 200pp. T.e.g. 8vo. Hinge in vol 2 starting else F. **$65.00\£36.00** *(JI 4/89)*

Stephens, Theo A. My Garden. London, 1954. 192pp. G in dj. **$18.00\£10.00** *(ABx 3/89)*

Stern, F.C. A Chalk Garden. London, 1974. New ed. 198pp. G in dj. **$21.50\£12.00** *(ABx 3/89)*

Steuart, Sir Henry. The Planter's Guide. Edinburgh, 1828. 1st ed. [xxii], 473pp. 5 engraved plates incl doubled-page frontis full contemp. calf blind tooled. Marbled edges. A Fine, clean copy. **$288.00\£160.00** *(ABx 3/89)*

Steudel, E.T. Nomenclator Botanicus. Stuttgart, 1841. 2nd (last) ed. 2 vols. New clth. Stmps on title-pge, half-title fxed. **$144.00\£80.00** *(WW 1/90)*

Stevens, G.A. Climbing Roses. NY, 1933. 220pp. VG. **$16.00\£9.00** *(AB 4/89)*

Stevens, G.A. Roses in the Little Garden. Boston, 1926. 118pp. VG. **$12.50\£7.00** *(RW 6/89)*

Stevens, William C. Kansas Wild Flowers. Lawrence, 1948. 463pp. Sm 4to, blue cloth. Author's inscription. **$22.50\£13.00** *(BB 12/88)*

Stevenson, J.A. An Account of Fungus Exsiccati Containing Material from the Americas. Beih. Nov. Hedw. 36, 1971. [viii], 564pp. Ry 8vo. Wrpps. G. **$63.00\£35.00** *(WW 1/90)*

Stevenson, J.B. (ed). The Species of Rhododendron. Np, 1947. 2nd ed. (1930). [vii], 861pp, [31]. Blue cloth. Fine. **$85.00\£47.00** *(BB 12/88)*

Stevenson, Violet. Patio, Rooftop and Balcony Gardening. London, 1967. 127pp. G in dj. **$9.00\£5.00** *(ABx 3/89)*

Stewart, Alban. A Botanical Survey of the Galapagos Islands. SF, 1911. 7-288pp. Wrappers, VG. **$38.00\£21.00** *(BB 12/88)*

Stewart, C. Topiary. Golden Cockerel Press, 1954. 34pp. Hlf bckram. Ltd to 500 cpies. G. **$72.00\£40.00** *(WW 1/90)*

Stewart, J.L. & D. Brandis. The Forest Flora of N.W. & Central India. 1874. 1/2 morocco, wrn; frnt cvr loose else G. **$36.00\£20.00** *(DL 89/90)*

Stockwell, L. & W. Breazeale. Arizona Cacti. AZ, 1933. 116pp. Wrs, VG. **$13.00\£7.00** *(GW 12/88)*

Stodola, Jiri. Encyclopedia of Water Plants. Jersey City, 1967. 368pp. Illus. glossy boards. Fine. **$17.00\£9.00** *(BB 12/88)*

Stoker, Fred. A Gardener's Progress. NY, 1939. 457pp. VG in well worn dj. **$15.00\£8.00** *(AB 4/89)*

Stone, B.C. The Genus Pelea A. Gray. (Rutaceae: Evodiinae). A Taxonomic Monograph. London, 1969. 188pp. Wrps. G. **$21.00\£12.00** *(WW 1/90)*

Stone, Witmer. The Plants of Southern New Jersey. Trenton, 1911. 828pp. F. **$40.00\£22.00** *(WH 1/90)*

Stones, M. & W.M. Curtis. The Endemic Flora of Tasmania. London, 1967. Vol I. 70pp. Folio. Clth. G. **$54.00\£30.00** *(WW 1/90)*

Stout, A.B. Daylilies. NY, 1934. [x], 119pp. Head/foot of spine rubbd. Ownr's sig else F. Quite scarce. **$90.00\£50.00** *(Hrt 12/89)*

Stracke, J. Richard (ed). The Laud Herbal Glossary. Amsterdam, 1974. Wraps. As New. **$33.00\£18.00** *(Hrt 12/89)*

Street, Frederick. Rhododendrons. London, 1965. 177pp. 7x10. G. **$32.50\£18.00** *(ABx 3/89)*

Strettell, G. The Ficus Elastica in Burma Propee Proper. Rangoon, 1876. 222pp. Several pages are photo copied. Bound. VG. **$15.00\£8.00** *(GW 8/89)*

Stroud, Dorothy. Capability Brown. London, 1975. 262pp. 1st prntng of this ed. F in dj. **$48.00\£27.00** *(BF 2/90)*

Strutt, Jacob George. Sylva Britannica, or Portrait of Forest Trees. London, 1830. [xvi], 151pp. Mnr fxng, chppd title, else VG in orig bndng. **$340.00\£188.00** *(Hrt 12/89)*

Stuart, C.M. Villiers. Spanish Gardens, Their History, Types and Features. 1929. 1st ed. Scarce. Nice cpy in dj. **$63.00\£35.00** *(DL 89/90)*

Stuart, George. The Scottish Gardener. London, 1963. 288pp. G. **$21.50\£12.00** *(ABx 3/89)*

Subik, Rudolf. Decorative Cacti. London, 1973. 132pp. VG. **$20.00\£11.00** *(Hrt 12/89)*

Another copy. London, 1973. 132pp. F in chppd dj. **$15.00\£8.00** *(CB 1/90)*

Sudell, Richard. Landscape Gardening. London, 1948. 5th imp. 480pp. G. **$32.50\£18.00** *(ABx 3/89)*

Sudworth, G. Forest Trees of the Pacific Slope. USDA, 1908. 441pp. Bound. VG. **$15.00\£8.00** *(GW 8/89)*

Sullivant, William S. Icones Muscorium, or Figures and Descriptions of Most of Those Mosses Peculiar to Eastern North America Which Have Not Been Heretofore Figured. (author). Cambridge, Mass., 1864. Ry 8vo. [viii], 216pp. Bnd w/very rare supplement. W/same title except for By the Late (author) Supplement, Posthumous. Cambridge, 1874. 77 copper plts (81 called for/mssng 10-11-46-57). Appear not to have been bnd in originally. [viii], 109pp. Green libry bndng, lettering frnt/spine, reinforced bndng, lght fxng to a number of plts. Very nice cpy. **$900.00\£500.00** *(Hrt 12/89)*

Summerhayes, V.S. Wild Orchids of Britain. New Naturalist, 1951. [xvii], 366pp. 1st ed. G. **$45.00\£25.00** *(WW 1/90)*

Sunding, P. The Vegatation of Gran Canaria. Oslo, 1972. 186pp. Wrs. **$20.00\£11.00** *(GW 8/89)*

Sung, T. Weeds Found on Cultivated Land in Western Taiwan. China, 1964. 185pp. VG. **$20.00\£11.00** *(GW 8/89)*

Sutton & Sons. Culture of Vegetables & Flowers from Seeds & Roots. London, 1930. 18th ed. 462pp. As New. **$15.00\£8.00** *(AB 2/89)*

Sutton, L.N. The Cool Greenhouse. London, 1935. Reprint. 186pp. G. **$10.75\£6.00** *(ABx 3/89)*

Suzuki, S. Roses of the World. Kanda-Tokyo, 1956. 4to. Dedic. to E.S. Boerner. F. **$45.00\£25.00** *(PBx 11/89)*

Swanton, E.W. Fungi and How to Know Them. 1909. 1st ed. VG in dj. **$10.50\£6.00** *(DL 89/90)*

Swinson, Arthur. Frederick Sander: The Orchid King. London, 1970. 252pp. G in dj. **$36.00\£20.00** *(ABx 3/89)*

Swinson, Arthur. Frederick Sanders: The Orchid King. London, 1970. 252pp. Nr F in dj. **$40.00\£22.00** *(Hrt 12/89)*

Swinstead, G. & R.I. Hillyard. The Story of My Old-World Garden. London, 1910. 4to. 55pp. Decorated cloth. G. **$72.00\£40.00** *(ABx 3/89)*

Symons, Les. Agricultural Geography. NY. Praeger, 1967. 283pp. VG in worn dj. **$10.00\£6.00** *(AB 4/89)*

Symons-Jeune, B. Natural Rock Gardening. Country, 1932. 151pp. VG. **$12.50\£7.00** *(GW 12/88)*

Symons-Jeune, B.H.B. Natural Rock Gardening. London, 1936. 2nd ed. Trifle used. **$21.00\£12.00** *(WW 1/90)*

Synge, P. Journal of the Royal Horticultural Society. 1958. 568pp. VG. **$12.75\£7.00** *(GW 12/88)*

Synge, P.M. A Diversity of Plants. London, 1953. G. **$21.00\£12.00** *(WW 1/90)*

Synge, P.M. Collin's Guide to Bulbs. 1971. 2nd ed. VG in dj. **$27.00\£15.00** *(DL 89/90)*

Synge, P.M. In Search of Flowers. London, 1973. 176pp. Ry 8vo. F in dj. **$36.00\£20.00** *(WW 1/90)*

Synge, Patrick. In Search of Flowers. London, 1973. 176pp. G in dj. **$21.50\£12.00** *(ABx 3/89)*
Another copy. London, 1973. 1st ed. 176pp. VG in dj. **$30.00\£17.00** *(JI 12/88)*

T

Tabor, Grace. Come into the Garden. NY, 1921. 324pp. Light wear, hinges repaired. **$10.00\£6.00** *(AB 2/89)*
Another copy. NY, 1921. 324pp. VG. **$20.00\£11.00** *(RW 6/89)*

Tabor, Grace. Old Fashioned Gardening. NY, 1913. [ix], 263pp. Frnt hnge crackd; ownr's inscrp f.f.e.p. Lght stn to frnt cvr. G. **$30.00\£17.00** *(CB 1/90)*

Tabor, Grace. Old Fashioned Gardening. NY, 1913. 263pp. Cloth worn, inner hinges repaired. **$18.00\£10.00** *(AB 2/89)*

Tabor, Grace. The Landscape Gardening Book. NY, 1911. 1st, crown 4to, [xii], 180pp. Glt stmpd pict clth. VG. **$45.00\£25.00** *(Hrt 12/89)*

Tabor, Grace. The Landscape Gardening Book. NY, 1911. 1st ed. 180pp. VG. **$35.00\£19.00** *(RW 6/89)*

Tackholm, V. et al. Students' Flora of Egypt. Cairo, 1956. 649pp. Ry 8vo. G. **$81.00\£45.00** *(WW 1/90)*

Taft, L.R. Greenhouse Construction. NY, 1914. 210pp. VG. **$15.00\£8.00** *(AB 4/89)*

Tagawa, M. Coloured Illustrations of the Japanese Pteridophyta. Osaka, 1959. G. **$36.00\£20.00** *(WW 1/90)*

Takhtadjan, A.L. Flora Arminii. Erevan, 1954-73. 6 vols. In Russian. Clth. G. **$108.00\£60.00** *(WW 1/90)*

Tamura, Tsuyoshi. Art of the Landscape Gardens in Japan. Tokyo, 1935. [xx], 245pp. Intro & 1st 90pp on uncut, possibly hand-made paper. Balance on coated paper in VG cond. Some spine fading and tenderness. Box enclosure trn & faded. Ownr stmp inside frnt cvr else overall VG. **$130.00\£72.00** *(Hrt 12/89)*

Tansley, A.G. Britain's Green Mantle. London, 1949. [xii], 294pp. G. **$21.00\£12.00** *(WW 1/90)*

Tansley, A.G. The British Islands and Their Vegetation. Cambridge, (1949), 1965. 970pp. G. **$99.00\£55.00** *(WW 1/90)*

Tatsui, Matsunosuke. Japanese Gardens. Tokyo, 1936. 2nd ed. Tourist Library #4. 111pp. Wrps. G. **$22.00\£12.00** *(Hrt 12/89)*

Taubenhaus, J.J. Culture and Diseases of the Sweet Potato. NY, 1923. 286pp. Light wear. Scarce. **$25.00\£14.00** *(AB 2/89)*

Taylor, Albert D. The Complete Garden. NY, 1921. [xxviii], 440pp. VG. **$25.00\£14.00** *(RW 6/89)*

Taylor, G.C. Garden Making by Example. London, 1932. 1st ed. 130pp. G. **$20.00\£11.00** *(Hrt 12/89)*

Taylor, Geoffrey. Some Nineteenth Century Gardeners. London, 1951. 175pp. Small 8vo. G in worn dj. **$21.50\£12.00** *(ABx 3/89)*

Taylor, Geoffrey. The Victorian Flower Garden. London, 1952. 212pp. Small 8vo. G. **$21.50\£12.00** *(ABx 3/89)*
Another copy. London, 1952. 1st ed. 212pp. 12mo. Green cloth. **$17.50\£10.00** *(BB 12/88)*

Taylor, George. An Account of the Genus Meconopsis. London, 1934. 1st ed. [xiii], 130pp. Spine fade else VG. Scarce. **$85.00\£47.00** *(Hrt 12/89)*
Another copy. London, 1934. 144pp. G. **$27.00\£15.00** *(WW 1/90)*

Taylor, George. An Account of the Genus Meconopsis. London, 1934. 1st ed. [xiv], 130pp. Faded spine else VG. **$45.00\£25.00** *(JI 12/88)*

Taylor, Gladys. Old London Gardens. London, 1953. 1st ed. 194pp. VG in dj w/small tear. **$30.00\£17.00** *(JI 7/88)* *Another copy.* London, 1953. 1st ed. 194pp. G. **$25.00\£14.00** *(CB 8/89)*

Taylor, Kathryn S. & Stephen F. Hamblin. Handbook of Wild Flower Cultivation. NY, (1963), 1973. [xi], 307pp. Fine in torn dj. **$14.50\£8.00** *(BB 12/88)*

Taylor, Norman (ed). Taylor's Encyclopedia of Gardening, Horticulture and Landscape Design. Boston, 1948. 2nd ed. Rev. & enl. [xii], 1225pp. VG. **$10.00\£6.00** *(RW 6/89)*

Taylor, Norman. Color in the Garden & Fragrance in the Garden. NY, 1953. 118pp + 110pp. VG in dj. **$10.00\£6.00** *(RW 11/86)*

Taylor, Norman. Wild Flower Gardening. Princeton, 1955. [xiii], 128pp. G. **$18.00\£10.00** *(Hrt 12/89)*

Taylor, Raymond L. Plants of Colonial Days. 1952. 107pp. Wraps, VG. **$4.00\£2.00** *(RW 11/86)*

Terres, John K. Songbirds in Your Garden. NY, 1953. [xiv], 274pp. G in dj. **$21.50\£12.00** *(ABx 3/89)*

Tharp, B.C. and Chester V. Kielman (eds). Mary S. Young's Journal of Botanical Explorations in Trans-Pecos, Texas. August-September, 1914. Austin, The Southwestern Historical Quarterly. Jan, 1962 (366-393pp) & April, 1962 (512-538pp). Vol LXV, nos. 3 & 4. VG. **$55.00\£31.00** *(JI 4/89)*

Tharp, Louise Hall. Saint-Gaudens and the Gilded Era. Boston, 1969. [xii], 419pp. VG in sl wrn dj. **$35.00\£19.00** *(Hrt 12/89)*

The Architectural Record. NY, Arch. Recd. Jan., Feb. and Nov., 1911. Pictorial wraps, covers loose, spines chipped. **$20.00\£11.00** *(MD 11/88)*

The Complete Gardener and Florist. NY, 1848. 8th ed. Bound w/ The Flower Gardener. NY, 1850. Together 144pp. Decor prntd brds respined. Sound and Clean. **$35.00\£19.00** *(PBx 11/89)*

The Garden, An Illustrated Journal of Horticulture in All Its Branches. London. Vol. XIV, Christmas, 1878. [xii], 588pp. Some fxng affecting only a few plts. Rear hnge open, covers are flaked and soiled. **$75.00\£42.00** *(BF 2/90)*

The Garden, An Illustrated Weekly Journal of Horticulture in All Its Branches. London. Vol. #74, 1910. 3/4 green morocco, rubbed; contents VG. **$75.00\£42.00** *(BF 2/90)*

The Garden Club of America. History 1913-1938. Phil, nd. 177pp. VG. **$30.00\£17.00** *(Hrt 12/89)*

The Horticulturist and Journal of Rural Art and Rural Taste. NY, 1864 & 1865. 2 vols. 384 & 396pp. 8vo. Blindstamped green cloth. Cover of 2nd volume has some staining. **$120.00\£67.00** *(MD 11/88)*

The Ideal Garden Book. London, n.d. (1930's). [x], 252pp. VG in chipped dj. **$9.00\£5.00** *(CB 12/89)*

The Oxford Book of Garden Flowers. Oxford, 1963. Sq. 8vo. [viii], 207pp. VG in dj. **$10.00\£6.00** *(RW 11/86)*

The Shepherd's Kalendar. London, n.d. (circa 1710). 4th ed. [vi], 157, (3). New clf, antique style. Few marginal defects and stning. G. **$216.00\£120.00** *(WW 1/90)*

The Spirit of the Woods. London, 1837. Ry 8vo. 306pp. F in modern binding. **$245.00\£136.00** *(PBx 11/89)*

Theophrastus. De Historia, et Causis Plantarum. Paris, 1529. [xxxvi], 343pp, [25]. 12mo, new clf. Antique style. Some marginal worming and light waterstaining not affecting the text. **$396.00\£220.00** *(WW 1/90)*

Thomas, Dr. G.L., Jr. Garden Pools, Water-Lilies and Goldfish. Princeton, 1958. 1st ed. [xviii], 222pp. VG in chipped dj. **$15.00\£8.00** *(JI 7/88)*

Thomas, G.S. The Old Shrub Roses. London, 1957. G. **$21.00\£12.00** *(WW 1/90)*

Thomas, George C., Jr. Roses For All American Climates. NY & London, 1924. [xii], 251pp. VG. **$17.00\£9.00** *(CB 1/90)*

Thomas, George C. Jr. The Practical Book of Outdoor Rose Growing. Philadelphia, 1920. 224pp. 5th ed. Lt green cloth, gold stamping. Color onlay. Soiled, front hinge starting else VG. **$45.00\£25.00** *(CB 8/89)*

Thomas, Graham Stuart. Colour in the Winter Garden. London, 1968. Garden Book Club. [xi], 220pp. VG in dj. **$15.00\£8.00** *(RW 6/89)*

Thomas, H.H. Round the Year in the Garden. NY, 1916. [xii], 275pp. VG. **$20.00\£11.00** *(RW 6/89)*

Thomas, H.H. The Rose Book. London, 1913. 1st ed. 287pp. Some marginal pencil notes. Light crease to inside corner of some plates else VG. **$22.00\£12.00** *(AB 4/89)*

Thomas, J.J. The American Fruit Culturist. 1920. 21st ed. 823pp. F. **$65.00\£36.00** *(PBx 11/89)*

Thomas, John H. Flora of the Santa Cruz Mountains of California. Stanford, 1961. [vi], 434pp. Blue cloth. New in dj. **$19.95\£11.00** *(BB 12/88)*

Thomas, Rose Haig. Stone Gardens. London, 1905. 14 fll pp color, (#3 has extra b\w plt), two photos, 12pp & short text for each design. Orig pict paper brd showing frequent use, interior reasonably G folio. **$95.00\£53.00** *(Hrt 12/89)*

Thomas, Sir William Beach. Gardens. London, 1952. [xviii], 302pp. VG. **$15.00\£8.00** *(RW 6/89)*

Thomas, W. & A. Collett. The English Year - Summer. London, 1920. 341pp. Cloth worn, contents VG. **$20.00\£11.00** *(AB 2/89)*

Thomassin, Simon. Recueil des Figures, Groupes, Thermes, Fontaines, Vases, Statues, et Autres Ornemens de Versailles. Se vend a Amsterdam, Chez Pierre Mortier, Librairie sur le Vygen-Dam, 1695. 4 vols in 1. 59pp of preliminaries. Vol I: title pp, blank, large fldng map of Versailles, 53 plts. Vol II: title pp, blank, plts 54-85. Vol III: blank, plts 86-172. Vol IV: title pp, blank, 173-218. Lght fxing, sm tears to crnrs, else plates clean. Some smudging to edges, frontis. Full contemp lthr bndng wrn, 5 diapers & title. Else VG. **$925.00\£514.00** *(Hrt 12/89)*

Thompson, C.J.S. The Mystery and Lure of Perfume. Phil, 1927. [xvi], 247pp. G. **$40.00\£22.00** *(Hrt 12/89)*

Thompson, C.R. Pruning Apple Trees. London, 1966. 271pp. Color frontis. VG in dj. **$18.50\£10.00** *(BB 12/88)*

Thompson, H. Stuart. Flowering Plants of the Riviera. London, 1914. [xxix], 249pp. VG. **$49.00\£27.00** *(BB 12/88)* *Another copy.* London, 1914. [xxx], 249pp. G. **$27.00\£15.00** *(ABx 3/89)*

Thompson, H.S. Sub-Alpine Plants. London, 1912. [xv], 325pp. G. **$27.00\£15.00** *(WW 1/90)*

Thompson, J.V. A Catalogue of Plants Growing in the Vicinity of Berwick upon Tweed. London, 1807. [xxiv], [1], 132pp. Stmp on title-pge. G. **$144.00\£80.00** *(WW 1/90)*

Thompson, Robert. The Gardener's Assistant. Glasgow, Edinburgh & London, 1859. [xv], 774pp. Half green morocco with marbled boards, these slightly rubbed, and edges 10x7. VG contemp. binding. **$198.00\£110.00** *(ABx 3/89)*

Thonner, F. The Flowering Plants of Africa. London, 1916, 1962. [xvi], 647pp. G. **$117.00\£65.00** *(WW 1/90)*

Thornber, J.J. & F. Bonker. The Fantastic Clan. NY, 1932. 194pp. Scarce. G. **$15.00\£8.00** *(WH 1/90)*

Thornton, R.J. The Temple of Flora. London, 1972. 66pp. G in dj. 4to. **$27.00\£15.00** *(ABx 3/89)*

Thorpe, Carlyle. A Journey to the Walnut Sections of Europe and Asia. LA, priv prntd. 1923. 101pp. Slght shlfwr, else VG in 1/4 clth cvrd brds. **$24.00\£13.00** *(Hrt 12/89)*

Thorpe, Francis Newton. An American Fruit Farm. NY, 1915. [xii], 348pp. VG. **$12.50\£7.00** *(RW 6/89)*

Tilden, J. Minnesota Algae. Vol. I. MN, 1910. 328pp. VG. **$20.00\£11.00** *(GW 12/88)*

Tipping, H. Avray. Gardens Old and New. Vol. I. London, nd (c. 1910). [xxiv], 299pp. VG. **$75.00\£42.00** *(RW 11/86)*

Tjomsland, Anne (M.D.). The White Potato. CIBA Symposia vol. II #6. 1950. 1254-1284pp. Wrps. VG. **$12.00\£7.00** *(Hrt 12/89)*

Todd, S. American Wheat Culturist. NY, 1868. 432pp. VG/F. **$45.00\£25.00** *(AB 2/89)*

Tolle, Leon J. (Jr.). Floral Art for Religious Events. NY, 1969. 192pp. VG in dj. **$20.00\£11.00** *(Hrt 12/89)*

Torfrida. Flowering Trees of India. Wellington, Nilgiris, 1944. 41pp. VG. **$12.00\£7.00** *(BB 12/88)*

Torfrida. Wild Flowers of India. Wellington, Nilgiris, 1944. 45pp. Boards, VG. **$12.00\£7.00** *(BB 12/88)*

Torrey, J. A Flora of the State of New York. Vol. 1 (of 2). 1843. Folio, orig gilt decor bndng, 72 full page lithos. Mod rbbd; contents sound & tight. **$150.00\£83.00** *(PBx 11/89)*

Tosco, U. & A. Fanelli. Mushrooms & Toadstools. Crescent, 1973. 4to. 64pp. G. **$6.00\£3.00** *(WH 1/90)*

Touring Club Italiano. Conosi L'Italia. 1958. Vol. 2, La Flora. Orig brds. Nice clean cpy. **$27.00\£15.00** *(DL 89/90)*

Traill, C.P. Studies of Plant Life in Canada. Toronto, 1906. [xvii], 227pp. New and rev. ed. Orig clth, G. **$63.00\£35.00** *(WW 1/90)*

Traub, Hamilton P. The Amaryllis Manual. NY, 1958. 338pp. G. **$21.50\£12.00** *(ABx 3/89)*

Trautvetter, E.R.A. Conspectus Florae Insulurum Nowaja - Semlja. Ca. 1870s. In Latin. 46pp. Gilt decorated, flexible green cloth. Scarce. **$32.00\£18.00** *(BB 12/88)*

Treat, Mary. Injurious Insects of the Farm & Garden. NY, O.J., 1882. 288pp. Gilt, decorated cloth and contents As New. **$40.00\£22.00** *(AB 4/89)*

Tricker, William. The Water Garden. NY, 1897. 120pp. Decorative cloth. VG. **$55.00\£31.00** *(RW 11/86)*

Tropical Woods. Numbers 98, 100-106. Yale, 1957. Wrs. VG. **$15.00\£8.00** *(GW 12/88)*

Truex, Philip. The City Gardener. NY, 1964. [xxv], 344, [xi]pp. **$10.00\£6.00** *(RW 11/86)*

Tubeuf, K. von. Monographie der Mistel. Munich, 1923. [xii], 832pp. Buckram. G. **$72.00\£40.00** *(WW 1/90)*

Tukey, Harold B. Dwarfed Fruit Trees. NY, 1964. [xi], 562pp. As new in dj. **$45.00\£25.00** *(Hrt 12/89)*

Tulasne, L.R. & C. Selecta Fungorum Carpologia. Oxford, 1931. 3 vols. Facs reprnt. Orig clth, slightly rbbd. Scarce. **$270.00\£150.00** *(DL 89/90)*

Turner, William. Libellus de re Herbaria, 1538; The Names of Herbes, 1548. Ray Soc'y, 1965. Facs printed on Basingwerk parchment paper. F in dj. **$48.00\£27.00** *(PBx 11/89)*

Turner, Wm. Fruits and Vegetables under Glass. NY, 1912. 255pp. VG. **$25.00\£14.00** *(AB 2/89)*

Turrill, W.B. The Royal Botanic Gardens, Kew. London, 1959. 256pp. G. **$32.00\£18.00** *(WW 1/90)*

Tweedie, Mrs. Alec. Hyde Park Its History and Romance. NY, 1908. [xii], 383pp. T.e.g. Gilt stmpd clth, slightly tender else VG. **$55.00\£31.00** *(Hrt 12/89)*

Tyler, G. Directions for a Kitchen Garden. – Every Month of the Year. 1970. In labelled casette. Ltd ed of 1000. Sgnd. F. **$16.00\£9.00** *(PBx 11/89)*

U

Ucko, P.J. & G.W. Dimbleby (eds). The Domestication and Exploitation of Plants and Animals. London, 1969. [xxvi], 581pp. G. **$63.00\£35.00** *(WW 1/90)*

Ukers, W. All About Coffee. 1922. Trade journal. 796pp. Rubbed else VG. **$35.00\£19.00** *(GW 8/89)*

Ukers, William H. All About Coffee. 1935. 2nd ed. Very scarce. VG. **$117.00\£65.00** *(DL 89/90)*

Ulmann, M. Wertvolle Kautschukpflanzen des Gemassigten Klimas. Berlin, 1951. 562pp. Ry 8vo, brds. G. **$27.00\£15.00** *(WW 1/90)*

Underhill, Terry. Heaths and Heathers. London, 1971. 1st ed. 256pp. G in dj. **$27.00\£15.00** *(ABx 3/89)*

Underwood, Loring. The Garden and Its Accessories. Boston, 1906. [xiv], 215pp. Decorative cover. **$20.00\£11.00** *(RW 11/86)*

Underwood, Mrs. Desmond. Grey and Silver Plants. Newton Abbot. 1972 BCE (1971). 143pp. Scarce. Fine in dj. **$34.00\£19.00** *(BB 12/88)*
Another copy. London, 1971. 1st ed. 143pp. G in dj. **$39.50\£22.00** *(ABx 3/89)*

United Fruit Co. Conquest of the Tropics. NY, 1914. [xii], 368pp. Slt spotting front brd else quite G. **$34.00\£19.00** *(Hrt 12/89)*

United States Department of Agriculture. Reprt of the Comm. of Agric. 1864. 676pp. Scuffed but sound. Very clean inside. **$15.00\£8.00** *(PBx 11/89)*

United States National Herbarium, 28 (2). Smith, A.C. The American Species of Thibaudieae. 1932. 240pp. Pbnd. Unopened. G. **$20.00\£11.00** *(PBx 11/89)*

United States National Herbarium, 37 (5). Raven, P.H. A Revision of the Genus Camissonia (Onagraceae). 1969. 237pp. Pbnd. G. **$14.00\£8.00** *(PBx 11/89)*

United States National Herbarium, Vol #27. Standley, P.C. Flora of the Panama Canal Zone. 1928. Scarce. Bnd in green clth. VG. **$63.00\£35.00** *(DL 89/90)*

United States Sugar Corp. Sugar and the Everglades. Clewiston, FL, 1939. 55pp. Wrps. VG. **$10.00\£6.00** *(Hrt 12/89)*

Unknown. A Picture of the Seasons. Dublin, 1818. 176pp. Leather spine, worn boards, hinges and contents tight. **$50.00\£28.00** *(AB 2/89)*

Urban, Abram. My Garden of Dreams. Philadelphia, 1913. Limited author's edition. 148pp. VG. **$10.00\£6.00** *(RW 11/86)*

Urquhart, Beryl Leslie. The Camelia. 1956. Vols I & II. Folio. VG set in djs. **$270.00\£150.00** *(DL 89/90)*

Urquhart, Beryl Leslie. The Rhododendron. Sharpthorne, Sussex, 1958. Two vols. Vol I: 40pp. Vol II: 36pp. Folios. Endpaper maps. F in VG dj. **$320.00\£178.00** *(Hrt 12/89)*
Another copy. Vols I & II. Scarce. VG in orig dj. **$360.00\£200.00** *(DL 89/90)*

USDA, Forest Service. Biology of Alder. Portland, 1968. 292pp. Softbound. **$14.00\£8.00** *(BB 12/88)*

USDA Yearbook. Soil. Wash, 1957. 784pp. G. **$17.50\£10.00** *(WH 1/90)*

USDA. Woody-Plant Seed Manual. WA, 1948. 416pp. **$25.00\£14.00** *(GW 8/89)*

USNH. Contributions from Vol. 26: Systematic Plant Studies. WA, 1940. Complete in 10 parts. Wraps, VG. **$15.00\£8.00** *(GW 8/89)*

USNH. Contributions from Vol. 36. WA, 1960. Complete in 6 parts. Wraps. VG. **$15.00\£8.00** *(GW 8/89)*

USNH. Vol. 16. 1912. Bnd, VG. **$25.00\£14.00** *(GW 12/88)*

USNH. Vol. 37. WA, 1962. Complete in 6 prts. Wrs. VG. **$15.00\£8.00** *(GW 12/88)*

V

Van de Boe, Louis. Planning and Planting Your Own Place. NY, 1938. [xviii], 290pp. VG. **$15.00\£8.00** *(Hrt 12/89)*

Van de Pass, Crispen. Hortus Floridus. London, 1928. Vol. I only. Ltd. ed. 211/500. About F. **$125.00\£69.00** *(Hrt 12/89)*

Van Deman, Esther Boise. The Atrium Vestae. WA, 1909. 47pp. Grey wrps, spine worn away. Minor water stain. **$25.00\£14.00** *(MD 2/90)*

Van der Piji, L. Principles of Dispersal in Higher Plants. Springer, 1969. 153pp. VG in dj. **$20.00\£11.00** *(GW 8/89)*

Van Dersal, William. Native Woody Plants of the U.S. Wash, DC, 1938. 362pp. Green cloth. **$24.00\£13.00** *(BB 12/88)*

Van Meter, R. Bush Fruit Production. NY, 1928. 123pp. VG. **$13.00\£7.00** *(AB 2/89)*

Van Pelt Wilson, Helen. The Complete Book of African Violets. NY, 1951. 247pp. G in dj. **$14.50\£8.00** *(ABx 3/89)*

Van Rensselaer, Mrs. Schuyler. Art Out-of-Doors. Phil, 1959. [ix], 174pp. VG. **$20.00\£11.00** *(Hrt 12/89)*

Van Steenis, C. Pacific Plant Areas. Vol. I. Manila, 1963. 297pp. Wraps detached but present. **$20.00\£11.00** *(GW 8/89)*

Vansittart, Dr. (ed). Certain Ancient Tracts Concerning the Management of Landed Property. London, 1767. [vi], 82, [viii], 120, [viii], 100pp. Contemp. clf, jnts broken. G. $180.00\£100.00 (WW 1/90)

Varille, M. Bonheur des Jardins. Lyons, 1944. [viii], 135pp. Ltd to 500 cpies. G. **$36.00\£20.00** *(WW 1/90)*

Vasey, G. Illustrations of North American Grasses. Washington, 1891-93. 2 vols. New cloth. Stmp on title page and rev. of pltes. **$144.00\£80.00** *(WW 1/90)*

Vaucher, J.P. Histoire Physiologique des Plants d'Europe. Paris, 1841. 4 vols. Clf-bcked brds, trifle used. Rare. **$90.00\£50.00** *(WW 1/90)*

Veitch, James & Sons. A Manual of the Coniferae. London, 1881. 1st ed. 350pp. Some folding and one or two illus. Lightly foxed else nice, bright copy. **$108.00\£60.00** *(ABx 3/89)*

Veitch, James Herbert. A Traveller's Notes. Chelsea, 1896. 219pp. Original cloth lightly marked, else a VG copy. **$216.00\£120.00** *(ABx 3/89)*

Veitch, R. & J.H. Simmonds. Pests and Diseases of Queensland Fruits and Vegetables. Brisbane, 1929. 198pp. Brds. G. **$45.00\£25.00** *(WW 1/90)*

Velley, T. Coloured Figures of Marine Plants. Bath, 1795. [ii], 9, 3-8, [18]pp. Folio. New brds. Latin text missing pp 1-2. 1 plt missing. G. **$144.00\£80.00** *(WW 1/90)*

Verdoorn, F. (ed). Manual of Pteridology. London, 1967. [xv], 640pp. G. **$32.00\£18.00** *(WW 1/90)*

Vester, Bertha Spaffor. Flowers of the Holy Land. Kansas City, 1962. 64pp. VG. **$5.00\£3.00** *(RW 11/86)*

Vick, Edward. Audels Gardeners & Growers Guide. NY, 1928. 4 vols. Light wear. Endpapers stained and worn, else VG. **$18.00\£10.00** *(AB 2/89)*

Viereck, L.A. & E.L. Little, Jr. Alaska Trees and Shrubs. USDA, Wash, 1972. [viii], 265pp. Boards, F. **$15.00\£8.00** *(WH 1/90)*

Villiers-Stuart, C.M. Spanish Gardens. NY, 1929. 8vo, decorated blue cloth. 139pp. Modest wear. **$45.00\£25.00** *(MD 7/89)*

(von Humboldt). Intro by Bayard Taylor. The Life, Travels and Books of Alexander von Humboldt. NY, 1859. [xx], 482pp. (6) advts. mssng f.f.e.p. Needs rebndng, else reasonably G. **$21.00\£12.00** *(Hrt 12/89)*

von Humboldt, Alexander. Cosmos. London, 1871, 76, 78. 5 vols. Stain frnt cvr & 1st few pp of vol I. Contents VG in orig red clth. **$85.00\£47.00** *(Hrt 12/89)*

Von Miklos, Josephine. History, Beauty and Riches of Gardener's World. NY, 1969. 252pp. VG in dj. **$10.00\£6.00** *(AB 4/89)*
Another copy. VG in dj. **$20.00\£11.00** *(RW 11/86)*

Vyvyan, C.C. Letters from a Cornish Garden. London, 1972. 190pp. G in dj. **$21.50\£12.00** *(ABx 3/89)*

W

Wade, A.E. The Flora of Monmouthshire. Cardiff, 1970. 236pp. G. **$27.00\£15.00** *(WW 1/90)*

Wakefield, E.M. & R.W.G. Dennis. Common British Fungi. London, 1950. [ix], 290pp. Ry 8vo. G. **$50.00\£28.00** *(WW 1/90)*

Wakefield, P. An Introduction to Botany. London, 1818. 8th ed. [xii], 187pp. Hlf clf, trifle rbbed. **$36.00\£20.00** *(WW 1/90)*

Walcott, Mary Vaux. Illustrations of North American Pitcherplants. Wash, 1935. 8pp of text & intro, 15 col plts w/single pg of text for each plt separate, 34pp of text (includes 10 maps). Loose as issued in qrtr lthr binder. Glt stmpd spine. Sm rbbng head/foot of spine else F. Folio. **$135.00\£75.00** *(Hrt 12/89)*

Walker, Marian. Dahlias for Every Garden. NY, 1954. 128pp. Sgnd by Walker. VG. **$16.00\£9.00** *(CB 1/90)*

Walker, R. The Flora of Oxfordshire and Its Contiguous Counties. 1833. Some fxng to plts else a G cpy in orig clth. **$90.00\£50.00** *(DL 89/90)*

Walker, W. All the Plants of the Bible. Harper, 1943. 244pp. VG in worn dj. **$16.00\£9.00** *(GW 12/88)*

Walker, Winifred. All the Plants of the Bible. NY, 1957. 244pp. VG in chppd dj. **$14.00\£8.00** *(CB 1/90)*

Walling, Edna. Gardens in Australia. Melbourne, 1946. 4th prntng. 148pp. Blue clth, VG in worn dj. **$20.00\£11.00** *(MD 2/90)*

Walter, H. Die Vegetation der Erde. Band 2. Jena, 1968. 1001pp. Clth. G. **$32.00\£18.00** *(WW 1/90)*

Walther, Eric. A Key to the Species of Eucalyptus Grown in California. SF, 1928. Proceedings of the C.A.S., Vol. XVII, no. 3. 67-87pp. Gray wrappers. **$12.50\£7.00** *(BB 12/88)*

Walton, J. An Introduction to the Study of Fossil Plants. Adam, 1953. 201pp. VG in dj. **$25.00\£14.00** *(GW 8/89)*

Wang, Jen Yu & Gerald L. Barger. Bibliography of Agricultural Meteorology. Madison, 1962. [xi], 673pp. Rbbd else VG in dj. **$18.00\£10.00** *(Hrt 12/89)*

Ward, Lester. Check List. Flora of Washington, D.C. and Vicinity. Wash, 1882. 1384 items, pp 148-207. Orig wrps, some nibble on edges else VG. **$22.00\£12.00** *(Hrt 12/89)*

Ward, N.B. On the Growth of Plants in Closely Glazed Cases. London, 1842. [vii], 95, [3]pp. Orig clth (neatly reprd). 2 lvs with marginal stn else very clean. **$126.00\£70.00** *(WW 1/90)*

Warming, E. and M. Vahl. Oecology of Plants. Oxford, 1909, 1925. [xii], 422pp. Ry 8vo. G. **$45.00\£25.00** *(WW 1/90)*

Warner, Charles D. My Summer in a Garden. Boston, 1898, 1870. 212pp. Decorated cloth. As New. **$10.00\£6.00** *(AB 2/89)*

Warner, Charles Dudley. My Summer in a Garden. Boston, 1871. [xii], 183pp. Orig green clth, gilt titles. Cvr somewhat wrn. **$15.00\£8.00** *(AT 3/89)*

Warner, Robert. Select Orchidaceous Plants. London, 1862-1865. Reprnt by Rare Eds. Ltd, 1975. Ltd ed 0209/1000. Np, glt stmpd lthr. Very nice. **$300.00\£167.00** *(Hrt 12/89)*

Washington, G. Letters from His Excellency George Washington to Arthur Young. Esq. F.R.S. and Sir John Sinclair, Bart. M.P. Alexandria, 1803. 128pp. Quarter clf. Text a little brwnd and edges slightly wrn. Last lf tipped in. G. **$360.00\£200.00** *(WW 1/90)*

Wasson, R. Gordon. Soma and the Fly-Agaric, Mr. Wasson's Rejoinder to Professor Brough. Cambridge, Nov. 1972. 58pp. Wrps. VG. **$65.00\£36.00** *(Hrt 12/89)*

Waterfield, Margaret. Flower Grouping in English, Scotch & Irish Gardens. London & NY, 1907. [xiii], 237pp. Decorated cloth a little marked else a bright, sound copy. 4to. **$108.00\£60.00** *(ABx 3/89)*

Waterfield, Margaret. Garden Colour. London, 1905. 2nd ed. 196pp. 4to. **$54.00\£30.00** *(ABx 3/89)*

Waterston, J.M. The Fungi of Bermuda. Hamilton, 1947. 305pp. Name on endpapr, corners bumped. G. **$36.00\£20.00** *(WW 1/90)*

Watson, Rosamund M. The Heart of a Garden. Philadelphia, nd. 162pp. VG. **$20.00\£11.00** *(RW 11/86)*

Watson, William. Rhododendrons & Azaleas. London & Edinburgh, nd (1910?). 116pp. Covers lightly marked. **$18.00\£10.00** *(ABx 3/89)*

Watt, G. The Pests and Blights of the Tea Plant. Calcutta, 1898. [iii], 468, [xvii]pp. Stmp on flyleaf & title pg. Hlf morocco. G. **$45.00\£25.00** *(WW 1/90)*

Watts & Watts. Vegetable Growing Business. NY, 1940. 549pp. As New. **$20.00\£11.00** *(AB 2/89)*

Waugh, F.A. Kemp's Landscape Gardening. NY, 1911. 4th ed. [xxii], 292pp. Ownr's sig in ink else VG. **$65.00\£36.00** *(Hrt 12/89)*

Another copy. NY, 1912. 4th ed. [xxii], 292pp. Green clth, gilt stmpng. Extremities wrn, some lght sptting to spine else VG. **$45.00\£25.00** *(CB 1/90)*

Waugh, F.A. Kemp's Landscape Gardening. NY, 1912. 4th ed. [xxii], 292pp. Owner's sig else VG in gilt green clth. **$60.00\£33.00** *(Hrt 12/89)*

Waugh, F.A. Landscape Gardening. NY, 1910. [viii], 152pp, [xiv] advts. Spine fade & rbbd else VG. **$30.00\£17.00** *(Hrt 12/89)*

Waugh, F.A. Landscape Gardening. NY, 1913. 2nd ed. rev. [xii], 156pp. **$16.00\£9.00** *(BB 12/88)*
Another copy. Same ed., rev., etc. VG. **$20.00\£11.00** *(RW 11/86)*

Waugh, F.A. Landscape Gardening. NY, 1915. 12mo. 2nd ed. [xii], 156pp. Some spine fade else VG. **$35.00\£19.00** *(Hrt 12/89)*

Waugh, F.A. Plums and Plum Culture. NY, 1901. 1st ed. [xx], 380pp. 12mo. Edges rubbed else VG. **$55.00\£31.00** *(JI 4/89)*

Waugh, Frank. Everybody's Garden. NY, 1930. 396pp. Pictorial brds. VG. **$35.00\£19.00** *(Hrt 12/89)*

Weathers, John (ed). Commercial Gardening. 4 vols. London, 1913. 239, 235, 239, 244pp. VG. **$55.00\£31.00** *(RW 11/86)*

Weatherwax, Paul. Indian Corn in Old America. NY, 1954. 1st ed. [xii], 258pp. VG. **$65.00\£36.00** *(JI 4/89)*

Weaver, Lawrence (ed). Small Country Houses of To-day. London, nd. (1910-11). [x], 224pp. T.e.g., spine fade & wr on brds else VG. **$150.00\£83.00** *(Hrt 12/89)*

Weaver, Lawrence. House and Gardens by E.L. Lutyens. London, 1913. 1st ed. Folio, [xi], 344pp. In new clth on brds. Orig external title plt on brds. **$275.00\£153.00** *(Hrt 12/89)*

Weaver, Lawrence. Houses and Gardens by E.L. Lutyens. London, 1913. [xl], 344pp. Cream buckram, spine slightly darkened & brown boards a little worn folio. **$270.00\£150.00** *(ABx 3/89)*

Webber, Ronald. Percy Cane, Garden Designer. Edinburgh, 1974. 192pp. Becoming scarce. As New in dj. **$40.00\£22.00** *(Hrt 12/89)*

Webster & Wilson. Agriculture in the Tropics. London, 1966. 488pp. Dj, contents As New. **$20.00\£11.00** *(AB 2/89)*

Webster, H.N. Herbs. Boston, 1947. New ed. G. **$21.00\£12.00** *(WW 1/90)*

Wedda, John. Gardens of the American South. NY, 1971. 256pp. VG in dj. **$15.00\£8.00** *(AB 2/89)*

Weed, Clarence M. Ten New England Blossoms and Their Insect Visitors. Boston, 1895. [xi], 142pp. VG. **$10.00\£6.00** *(RW 11/86)*

Weed, Clarence Moores. Ten New England Blossoms and Their Insect Visitors. Boston, 1895. [xi], 142pp. Dec clth, slght soil. **$18.00\£10.00** *(Hrt 12/89)*

Weiner, Mike. Earth Medicine - Earth Foods. NY, 1972. 214pp. Nearly New in dj. **$20.00\£11.00** *(AB 4/89)*

Weiss, Prf. F.E. Fifty Years' Growth of the Botanical Department of Owens College and the University of Manchester. Dec., 1929. 8pp. Sgnd by authr. Wrps. VG. **$10.00\£6.00** *(Hrt 12/89)*

Welch, Charles F.S.A. History of the Worshipful Company of Gardeners. London, 1900. 2nd ed. 52pp + 15 illus. Half green calf with buckram boards. G. **$36.00\£20.00** *(ABx 3/89)*

Wells, A. Laurence. Garden Ponds, Fish and Fountains. London, 1949. Rev ed. 157pp. VG in wrn dj. **$30.00\£17.00** *(Hrt 12/89)*

Wells, Luana. The Children's Garden. London, 1960. 134pp. G in dj. **$9.00\£5.00** *(ABx 3/89)*

Welsh, Stanley L. et al. Common Utah Plants. Provo, 1965. 2nd ed. [iii], 312pp. Green cloth. VG. **$19.00\£11.00** *(BB 12/88)*

Weniger, Del. Cacti of the Southwest. Austin, nd (1970). [xv], 249pp. VG in dj. **$15.00\£8.00** *(RW 11/86)*

Weschke, Carl. Growing Nuts in the North. St. Paul, 1953. 124pp. As New. **$18.00\£10.00** *(AB 2/89)*

West, W. & G.S. A Contribution to the Freshwater Algae of Ceylon. Trans. Linn. Soc., 1902. Wrpprs. G. **$18.00\£10.00** *(WW 1/90)*

Westcott, Cynthia. Gardener's Bug Book. Dblday, 1964. 3rd ed. 625pp. Nearly New in dj. **$12.00\£7.00** *(AB 4/89)*

Westermarck, Nils. Policies, Planning & Management for Agricultural Development. Oxford, I.A.A., 1971. 616pp. Cloth and contents As New. **$15.00\£8.00** *(AB 4/89)*

Wheeler, Candace. Content in a Garden. Boston, 1901. 209pp. VG. **$20.00\£11.00** *(RW 11/86)*

Wheeler, D. & W. Hill. Grassland Seeds: A Handbook. . . VNR, 1957. 734pp. VG. **$15.00\£8.00** *(GW 8/89)*

Wheeler, W.A. Forage and Pasture Crops. NY, 1950. [xi], 751pp. G only. **$12.00\£7.00** *(CB 1/90)*

Wherry, E. Wild Flower Guide: North East & Midland United States. Doubleday, 1948. 202pp. VG. **$12.00\£7.00** *(GW 8/89)*

Another copy. Garden City, 1948. 202pp. F. **$10.00\£6.00** *(WH 1/90)*

Wherry, E.T. The Genus Phlox. 1955. Morris Arboretum Monographs III. Signed by Wherry. F. **$28.00\£16.00** *(PBx 11/89)*

Wherry, Edgar T. Guide to Eastern Ferns. Lancaster, 1937. 16mo, 220pp. Signed by Sen. Desmod. VG. **$25.00\£14.00** *(WH 1/90)*

Wherry, Edgar T. Wild Flowers of Mount Desert Island, Maine. Mt. Desert, 1928. [iv], 164pp. **$18.00\£10.00** *(BB 12/88)*

Whiffen, Thomas. The North-West Amazons. London, 1915. [xvii], 319pp. Gilt lettered spine. VG. **$110.00\£61.00** *(Hrt 12/89)*

White, Edward A. American Orchid Culture. NY, 1947. 4th prntng. 276pp. 3rd ed. Shaken. Some fxng throughout else G. **$23.00\£13.00** *(CB 1/90)*

White, Francis B.W. The Flora of Perthshire. Ed. by James Trail. Edinburgh, 1898. 1st ed. [lix], 407pp. 8vo. G. **$93.50\£52.00** *(ABx 3/89)*

Another copy. Edinburgh, 1898. [lix], 407pp. 500 cpies prntd. G. **$81.00\£45.00** *(WW 1/90)*

White, Gilbert. Ed. by Walter Johnson. Gilbert White's Journal. NY, 1970. [xlviii], 463pp. Some shlf wr bottom brds else VG. **$35.00\£19.00** *(Hrt 12/89)*

White, Gilbert. Garden Kalendar. 1751-1771. London, 1975. [xxviii], 443pp, 11pp of notes. VG. **$15.00\£8.00** *(RW 11/86)*

Another copy. London, 1975. [xxvii], 221 double pages, 12 of notes. Marbled boards and buckram spine, leather label. **$54.00\£30.00** *(ABx 3/89)*

White, J.E. Grant. Garden Art and Architecture. London, 1968. 8vo, grey boards. 110pp. F in dj. **$22.50\£13.00** *(MD 7/89)*

White, John. Rural Architecture. Glasgow, 1845. Folio, 90 plts, [4], 61pp, blank, the plts. Evenly brwned text. Clth on brds, orig title label on spine. G. **$275.00\£153.00** *(Hrt 12/89)*

White, W.H. The Book of Orchids. John Lane, Bodley Head, 1911. 2nd ed. 100pp. G. **$16.25\£9.00** *(ABx 3/89)*

White, William. Gardening for the South, or How to Grow Vegetables and Fruits. NY, 1868. 3rd ed. 444pp. G. **$30.00\£17.00** *(Hrt 12/89)*

Whitehill, Walter Muir. Dunbarton Oaks. Harvard University Press, 1967. 147pp. G. **$50.50\£28.00** *(ABx 3/89)*

Whiting, John R. A Treasury of American Gardening. NY, 1955. 1st ed. Lg 8vo. 272pp. VG in dj. **$10.00\£6.00** *(RW 11/86)*

Whitman, Georgia. Within My Garden Walls. NY, 1939. 165pp. VG. **$12.50\£7.00** *(RW 11/86)*

Whittle, Tyler. Some Ancient Gentlemen. NY, 1966. [x], 244pp. Yellow cloth. VG in dj. **$24.00\£13.00** *(BB 12/88)*

Another copy. 244pp. Taplinger, 1966. Ragged dj. Some ink notes. **$11.00\£6.00** *(GW 8/89)*

Whittle, Tyler. The Plant Hunters. London, 1970. [xiii], 257pp. About F in dj. **$45.00\£25.00** *(Hrt 12/89)*

Whittle, Tyler. The Plant Hunters. London, 1970. 1st ed. [xiv], 258pp. VG in dj. **$45.00\£25.00** *(JI 12/88)*

Wickson, E.J. California Vegetables in Garden and Field. SF, 1917. 4th ed. 319pp. VG. **$18.00\£10.00** *(AB 2/89)*

Wickson, E.J. The California Vegetables in Garden & Field. SF, 1913. 3rd rev ed. 326pp. VG. **$25.00\£14.00** *(PBx 11/89)*

Wickson, Edward J. The California Fruits. SF, 1926. 10th edn, rev. 509pp. G. **$15.00\£8.00** *(WH 1/90)*

Wiegand, K. & A.J. Eames. The Flora of the Cayuga Lake Basin, New York. 1926. 491pp. Bndng fair. First 2pp repred else sound, clean. **$22.00\£12.00** *(PBx 11/89)*

Wiesner, J. von. Die Rohstoffe des Pflanzenreichs. Weinheim & Lehre, 1962-28. 5th ed. Lief. 1 to 3, 5 & 7, 1236pp. Wrpprs. G. **$90.00\£50.00** *(WW 1/90)*

Wilder, Louise B. My Garden. Garden City, 1932. 317pp. VG. **$15.00\£8.00** *(AB 4/89)*

Another copy. NY, 1932. New edition. 317pp. G in dj. **$36.00\£20.00** *(ABx 3/89)*

Wilder, Louise Beebe. Adventure in My Suburban Garden. NY, 1931. [xv], 250pp. VG. **$30.00\£17.00** *(Hrt 12/89)*

Wilder, Louise Beebe. Colour in My Garden. NY, 1927. (1918). [xv], 410pp. Gilt stmpd clth; nice cpy. **$36.00\£20.00** *(Hrt 12/89)*

Wilder, Louise Beebe. Pleasures and Problems of a Rock Garden. NY, 1937. [xii], 294pp. Water stn frnt brd, sl tender else VG. **$28.00\£16.00** *(Hrt 12/89)*

Wilder, Louise Beebe. Pleasures and Problems of a Rock Garden. NY, 1937. Large 8vo. [xii], 294pp. VG. **$20.00\£11.00** *(RW 11/86)*

Wilder, Louise Beebe. The Fragrant Path. NY, 1936. Reprint. 407pp. G in dj. **$43.25\£24.00** *(ABx 3/89)*

Wilder, Louise Beebe. The Garden in Color. NY, 1937. 327pp. VG in dj. **$17.50\£10.00** *(WH 1/90)*

Another copy. NY, 1937. 1st ed. 4to. 327pp. VG in repaired dj. **$12.50\£7.00** *(RW 11/86)*

Wiley, Leonard. Rare Wild Flowers of North America. Portland, 1968. 501pp. Green cloth. Ltd. to 1000 copies. Signed and numbered. As New in dj. **$35.00\£19.00** *(BB 12/88)*

Wilkie, David. Gentians. London, 1936. 1st ed. 187pp. G in dj. **$57.50\£32.00** *(ABx 3/89)*

Williams, B.S. Select Ferns and Lycopods. London, 1873. 2nd ed. [viii], 353pp. Orig clth, frontis creased. G. **$32.00\£18.00** *(WW 1/90)*

Williams, Benjamin Samuel. Choice Stove and Greenhouse Ornamental-Leaved Plants. London, 1876. 2nd ed. [x], 350pp, [i-xxi], ads. F. **$45.00\£25.00** *(Hrt 12/89)*

Williams, Benjamin Samuel. Choice Stove and Greenhouse Ornamental Plants. London, 1876. 2nd ed. [x], 350pp, [xxi] of advts. Spine fade, gilt pictorial brds. **$30.00\£17.00** *(Hrt 12/89)*

Another copy. London, 1876. 2nd ed. Clth, used and faded. **$27.00\£15.00** *(WW 1/89)*

Williams, Henry. Window Gardening. NY, 1874. 10th ed. 302pp. VG. Inner hinges repaired. **$18.00\£10.00** *(AB 2/89)*

Willis, J.C. A Dictionary of the Flowering Plants and Ferns. Cambridge, (1931), 1957. 6th ed. Spine trifle wrn. **$27.00\£15.00** *(WW 1/90)*

Willkomm, M. & J. Lange. Prodromus Florae Hispanicae. Stuttgart, 1870-80, 1893. 3 vols. & suppl. 4 vols in 3, rare orig prtng. Bndng of supplement not quite uniform. **$270.00\£150.00** *(WW 1/90)*

Wilson, Adelaide B. Flower Arrangements for Churches. NY, 1952. 192pp. VG in dj. **$18.00\£10.00** *(Hrt 12/89)*

Wilson, Albert. Gardeners all in California. Menlo Park, 1953. 1st edition. 254pp. Inscribed by author. Fine in dj. **$15.00\£8.00** *(BB 12/88)*

Wilson, Charles Morrow. The Tropics; World of Tomorrow. NY, 1951. 275pp. F in dj. **$10.00\£6.00** *(WH 1/90)*

Wilson, E.H. America's Greatest Garden. Boston, 1925. 123pp. VG. **$15.00\£8.00** *(RW 11/86)*

Another copy. Boston, 1929, 4th ed. 123pp. Slt rub, own sig else quite G. **$24.00\£13.00** *(Hrt 12/89)*

Wilson, E.H. The Lilies of Eastern Asia. London, 1925. 1st ed. 110pp. 4to G. **$126.00\£70.00** *(ABx 3/89)*

Wilson, Ernest H. A Naturalist in Western China. London, 1913. 2 vols. Vol I: [xxxvii], 251pp. Vol II: [xi], 220pp. F cpy w/some very lght pencil underlining. **$375.00\£208.00** *(Hrt 12/89)*

Wilson, Ernest H. America's Greatest Garden. Boston, 1925. 123pp. Clipping laid in else VG. **$18.00\£10.00** *(CB 1/90)*

Wilson, Ernest H. Aristocrats of the Garden. Boston, 1926. 1st ed. [xxv], 312pp. t.e.g. partiallly uncut. VG in dj and shipping box from publisher. **$45.00\£25.00** *(RW 11/86)*

Another copy. Boston, 1932. 2 vols. 312pp. & 288pp. G. **$54.00\£30.00** *(ABx 3/89)*

Another copy. Boston, 1926. t.e.g. 312pp. slt tear bttm spine else about F. **$36.00\£20.00** *(Hrt 12/89)*

Wilson, Ernest H. Aristocrats of the Garden. Boston, 1926. [xxv], 312pp. Orig green clts. gilt titles & decors. Minor edge wr, else VG. **$35.00\£19.00** *(AT 3/89)*

Wilson, Ernest H. Aristocrats of the Garden. Boston, 1932. Two vols. [xxv], 312 & 288pp. VG. **$45.00\£25.00** *(RW 11/86)*

Wilson, Ernest H. Aristocrats of the Garden. London, 1937. 1st English ed. [xxiv], 308pp. G. **$36.00\£20.00** *(WW 1/90)*

Wilson, Ernest H. Aristocrats of the Trees. Boston, 1930. [xxi], 279pp. Spine fade else VG in blue stmpd clth. **$90.00\£50.00** *(Hrt 12/89)*
Another copy. Boston, 1930. [xxi], 279pp. Orig clth. G. **$108.00\£60.00** *(WW 1/90)*

Wilson, Ernest H. China, Mother of Gardens. Boston, 1929. [x], 408pp. Ry 8vo. Extremely scarce. Orig clth, trifle used. E.p.s. defective, else G internally. **$360.00\£200.00** *(WW 1/90)*
Another copy. Boston, 1929. 408pp. T.e.g., signed ed. F. **$325.00\£181.00** *(Hrt 12/89)*

Wilson, Ernest H. More Aristocrats of the Garden. Boston, 1928. 1st ed. [xiv], 288pp. T.e.g. partially uncut. VG in dj & box. **$445.00\£25.00** *(RW 11/86)*
Another copy. Boston, 1928. 1st ed. T.e.g. [xiv], 288pp. **$45.00\£25.00** *(CB 8/89)*
Another copy. Boston, 1928. 1st ed. [xiv], 288pp. T.e.g. Ownr's stmp, some pencilling, shelfwear else VG. **$28.00\£16.00** *(Hrt 12/89)*

Wilson, Ernest H. Plant Hunting. Boston, 1927. 2 vols. Vol I: [xxix], 248pp. Vol II: [ix], 276pp. But for sm ink sig vol I, F set. Glt on embossed brds. **$375.00\£208.00** *(Hrt 12/89)*

Wilson, Ernest H. The Lilies of Eastern Asia. London, 1925. [xiv], 110pp. Orig clth, rear e.p. creased. A little minor dust-soiling else G cpy of the 1st issue. **$117.00\£65.00** *(WW 1/90)*

Wilson, Helen Van Pelt. Perennials Preferred. NY, 1945. 256pp. VG in dj. **$12.50\£7.00** *(RW 11/86)*

Wilson, Helen. The Fragrant Year. NY, 1967. 306pp. VG in dj. **$20.00\£11.00** *(AB 4/89)*
Another copy. [xiii], 306pp. VG. **$20.00\£11.00** *(RW 11/86)*
Another copy. NY, 1967. [xiii], 306pp. F in dj. **$24.00\£13.00** *(Hrt 12/89)*

Wilson, Mary G.W. Enchantment of Gardens. NY & Boston, nd (1900s?). 218pp. G. **$27.00\£15.00** *(ABx 3/89)*

Winkenwerder, Hugo & Elias T. Clark. A Manual of Exercises in Forest Mensuration. Seattle, Privately printed, 1915. 1st ed. [iv], 148pp. VG. **$25.00\£14.00** *(JI 12/88)*

Winkler, A.J. et al. General Viticulture. Berkeley, 1974. [xx], 710pp. Revised & enlarged ed. F. **$19.00\£11.00** *(BB 12/88)*

Winkler H. Untersuchungen Ueber Propfbastarde. Pt. I, 1912. 186pp. Wraps, in German. G. **$30.00\£17.00** *(PBx 11/89)*

(Wirt, Elizabeth Washington Gamble). Flora's Dictionary. Baltimore, 1832. Dble title (orig of 1830 in new wrps?), fll gilt morocco. Some scuffing. **$120.00\£67.00** *(Hrt 12/89)*

Wister, J.C. Bulbs for American Gardens. Boston, 1930. [xxv], 278pp. F. **$40.00\£22.00** *(Hrt 12/89)*

Witham Fogg, H.G. The Flower Arranger's Garden. London, 1974. 235pp. G in dj. **$14.50\£8.00** *(ABx 3/89)*

Wittering, W.O. Weeding in the Forest: A Work Study Approach. London, 1974. [xvi], 168pp. Wrps. VG. **$15.00\£8.00** *(Hrt 12/89)*

Wolgensinger, Bernard. Villas in the Sun. Rutland, 1971. 160pp. About F in dj. **$70.00\£39.00** *(Hrt 12/89)*

Woman's Club of Havana. Flowering Plants from Cuban Gardens. 1951. Scarce. VG. **$27.00\£15.00** *(DL 89/90)*

Wood, Denis. Practical Garden Design. London, 1976. 160pp. As New in dj. **$20.00\£11.00** *(Hrt 12/89)*

Wood, Horatio C., Jr. A Contribution to the History of the Fresh-Water Algae of North America. Wash, 1872. [ix], 262pp. 4to. Rebacked. F. **$125.00\£69.00** *(WH 1/90)*

Woodcock, H.B.D. & William T. Stern. Lilies of the World. London/NY, 1950. 1st ed. 431pp. Scattered light foxing else near F. **$59.00\£33.00** *(BB 12/88)*
Another copy. Country Life, 1950. 431pp. VG in worn dj. **$50.00\£28.00** *(GW 12/88)*

Woodcock, H.D. & J. Coutts. Lilies. London, 1936. 2nd ed. 242pp. G. **$32.50\£18.00** *(ABx 3/89)*

Woodforde, J. A Catalogue of the Indigenous Phenogamic Plants Growing in the Neighborhood of Edinburgh. Edinburgh, 1824. [xi], 86pp. 12mo. orig brds, trfle wrn. A few libr. stmps and a little fxing. Very scarce. **$90.00\£50.00** *(WW 1/90)*

Woodward, M. Gerard's Herbal: The Essence Thereof Distilled By Marcus Woodward. Howe, 1927. 303pp. Faded spine, else VG. **$25.00\£14.00** *(GW 8/89)*

Wooley, J.C. Farm Buildings. NY, 1946. 354pp. As New. **$16.00\£9.00** *(AB 2/89)*

Wooster, David F.R.H.S. ed. Alpine Plants. London, 1872. 1st ed. [xii], 152pp. Handsome contemp, half black morocco with gilt dec spine. Occas. very light spotting but plates nearly all very clea. **$270.00\£150.00** *(ABx 3/89)*

Wooster, David. Alpine Plants: Figures and Descriptions of Some of the Most Striking and Beautiful of the Alpine Flowers. 1874. 2nd ed. 2 vols, orig green clth. Slght wear at head/foot of spine else very well preservd set. Exceptionally clean. **$324.00\£180.00** *(DL 89/90)*

Wooton, E.O. & Standley, P.C. Flora of New Mexico. Weinheim, 1971. 794pp. Wrpprs. G. **$36.00\£20.00** *(WW 1/90)*

Worlidge, J. Systema Agriculturae. London, 1681. 3rd ed. [xxviii], 326, [6]pp. Folio. Old clf, unlettered. Lower outer crnrs of first few lves defective. Engraved title prtly colored by hand. **$432.00\£240.00** *(WW 1/90)*

Wright, Horace J. & Walter Wright. Beautiful Flowers and How to Grow Them. London, 1922. 402pp. 7x10. G. **$36.00\£20.00** *(ABx 3/89)*

Wright, Horace. The Fruit Grower's Guide. London, nd (1920s?). 2 vols. 682pp. G. **$108.00\£60.00** *(ABx 3/89)*

Wright, J. The Fruit-Growers Guide. London, 1924. New and rev. ed. 2 vols. Imp 8vo. Orig clth, cvrs trifle stned, very slght internal stning beginning vol 2. **$117.00\£65.00** *(WW 1/90)*

Wright, Richardson. Gardener's Day Book. Phila, 1938. 384pp. VG in worn dj. **$12.00\£7.00** *(AB 2/89)*

Wright, Richardson. Gardener's Tribute. Phil, 1949. 1st ed. 256pp. VG in older dj. **$18.00\£10.00** *(Hrt 12/89)*

Wright, Richardson. Gardener's Tribute. Phila, 1949. 256pp. VG in dj. **$15.00\£8.00** *(AB 2/89)*

Wright, Richardson. Gardening with the Experts. NY, 1941. 239pp. VG in worn dj. **$15.00\£8.00** *(AB 4/89)*

Wright, Richardson. Greedy Gardeners. Philadelphia & NY, 1955. 1st ed. 95pp. G. **$21.50\£12.00** *(ABx 3/89)*

Wright, Richardson. Small House and Large Garden. Boston, 1924. 219pp. VG. **$10.00/£6.00** *(AB 4/89)*

Wright, Richardson. Story of Gardening. G.C., 1938. 475pp. VG in lightly worn dj. **$20.00\£11.00** *(AB 4/89)*
Another copy. NY, 1934. 1st ed. [x], 475pp. VG. **$20.00\£11.00** *(RW 11/86)*

Another copy. London: Routledge, 1934. 475pp. G. **$39.50\£22.00** *(ABx 3/89)*

Wright, Richardson. The Gardener's Bed Book. Phila, 1929. 1st ed. 341pp. Light wear. From the library of May Watts, with her book plate. **$20.00\£11.00** *(AB 2/89)*

Wright, Richardson. The Gardener's Bed Book. Phila, 1929. 2nd ed. 341pp., cloth with light wear. **$10.00\£6.00** *(AB 4/89)*

Another copy. Philadelphia & London, 1929. 341pp. G. **$18.00\£10.00**

Wright, Richardson. The Story of Gardening. NY, 1935. [x], 475pp. VG in tattered dj. **$16.00\£9.00** *(CB 1/90)*

Wright, Richardson. Winter Diversions of a Gardener. Phila, 1934. 356pp. Spine sun faded. Else VG. **$10.00\£6.00** *(AB 4/89)*

Wright, W. An Account of the Medicinal Plants Growing in Jamaica. London, 1787. 79pp. Lib bkplte, no stmps. Clf (neatly rebckd). **$90.00\£50.00** *(WW 1/90)*

Wright, W.P. Alpine Flowers and Rock Gardens. London, 1910. 292pp. Ry 8vo, orig decor clth, corners bumped. A little foxing. G. **$45.00\£25.00** *(WW 1/90)*

Wright, Walter P. Garden Trees and Shrubs. London, 1928. Rev. ed. 355pp. VG. **$25.00\£14.00** *(RW 11/86)*

Wright, Walter P. Pictorial Practical Rose Growing. London, 1914. 152pp. G only. **$15.00\£8.00** *(CB 1/90)*

Wu, Y.C., K.K. Wong & S.M. Pong. Polypodiaceae Yaoshanensis. Canton, 1932. 372pp. Imp 8vo. Orig wrpprs, trifle wrn. Sm lib stmp on covr. Presentation cpy from Y.C. Wu. Very scarce. **$117.00\£65.00** *(WW 1/90)*

Wulff, E.V. Istorischeskaya Geografia Rastenii. Moscow, 1944. [xix], 546pp. Ry 8vo. Text in Russian. G. **$90.00\£50.00** *(WW 1/90)*

Wyman, D. Shrubs and Vines for American Gardens. NY, 1969. 2nd rev. ed., 640pp. G. **$27.00\£15.00** *(WW 1/90)*

Wyman, Donald. Shrubs and Vines for American Gardens. NY, 1956. VG in dj. **$15.00\£8.00** *(RW 11/86)*

Wyman, Donald. The Arboretums and Botanical Gardens of North America. American Association of Botanical Gardens & Arboretums (reprinted from Chronica Botanica Vol X), 1947. Paper covers, G. **$21.50\£12.00** *(ABx 3/89)*

Wyman, Donald. The Saturday Morning Garden. NY, 1962. 1st ed. 236pp. Light cvr spts. **$18.00\£10.00** *(Hrt 12/89)*

Wynne, Peter. Apples History, Folklore, Horticulture, Gastronomy. NY, 1975. 280pp. As New in dj. **$15.00\£8.00** *(AB 2/89)*

Y

Yashiroda, Kan. Handbook on Japanese Herbs. NY, 1968. 72pp. Lib. bnd. cloth, contents As New. **$4.50\£3.00** *(AB 4/89)*

Yates, Lucy H. The Gardener and the Cook. NY, 1913. 260pp. VG. **$10.00\£6.00** *(RW 11/86)*

Yoshimura, Y. & G.M. Halford. The Japanese Art of Miniature Trees and Landscapes. Rutland, 1960. 220pp. VG. **$12.50\£7.00** *(WH 1/90)*

Young, Andrew. A Retrospect of Flowers. London, 1950. 176pp. G in dj. **$9.00\£5.00** *(ABx 3/89)*

(Young Jr., William). Botanica Neglecta/William Young Jr. Ed. by Samuel Rhoads. Phil, 1916. 250 nmbrd and sgnd cpies. Unnumbered. [xi], (1), facsimile, (3), 55pp. Clth cvrd brds. F. **$145.00\£81.00** *(Hrt 12/89)*

Z

Zaugg, Hans & Hans Coaz. Decorative Trees and Shrubs. NY, 1960. 4to. 60 color plates. F in dj. **$30.00\£17.00** *(WH 1/90)*

Zeller, H. Wild Flowers of the Holy Land. London, 1883. 3rd ed. [xiv]pp. with 54 coloured pltes. Orig clth. F.f.e.p. removed, occ very minor fxing but G. **$153.00\£85.00** *(WW 1/90)*

Zenkert, Charles. The Flora of the Niagara Frontier Region. Buffalo, 1934. 329pp. Signed by author. F. **$37.50\£21.00** *(WH 1/90)*

Zocchi, Giuseppe. Vues de Florence et de Toscane. Geneve, 1974. 224pp. Wrps. VG. **$27.00\£15.00** *(Hrt 12/89)*

Zucker, Isabel. Flowering Shrubs. Princeton, 1966. [xx], 380pp. VG with dj. **$23.50\£13.00** *(BB 12/88)*

Key to Abbreviations & Glossary of Terms Used

abt. About

advs. Advertisements

a.e.g. All edges gilt (all edges have gold color applied)

bckd. Backed (covering the spine)

bckrm. Buckram (heavy binding fabric)

bkplte. Bookplate

bmpd. Bumped (corners somewhat damaged)

bnd. Bound

bndng. Binding (exterior covering)

brds. Boards (stiff material used for covers)

chppd. Chipped (nicked)

clth. Cloth

contemp. Contemporary

cpy. Copy (a book)

crckd. Cracked (hinges beginning to open)

crnr. Corner

cuts. Engravings

cvr. Cover

dec, decor. Decorative

dentelles. Fine, tooled scrollwork

diapers. Geometrical pattern on binding

diced. Pattern resembling cubes

dmp stning. Damp staining (slightly browning from moisture)

ed. Edition

enlrgd. Enlarged

e.p. End paper (sheets added to a book connecting the body to the binding; "free" end papers are not pasted)

facs. Facsimile

f.f.e.p. Front free end paper

frnt. Front

fxng. Foxing (light brown spots)

hlf. Half

hlf clf. Half calf (spine bound in leather; extending across boards)

illus. Illustrated

imp. Impression

inscr. Inscription

jnt. Joint (exterior junction of the spine and sides)

lf. Leaf

ltd. Limited

lthr. Leather

lves. Leaves

mdrn. Modern

mod. Moderately

mrbld. Marbled

mrgns. Margins

mrks. Marks

mssng. Missing

nd. No date

np. No place (place of publication)

nr. Near

orig. Original

pbl. Published

pict. Pictorial

plt. Plate

pp. Pages

prelims. Preliminaries (front matter; title page, dedication, table of contents, etc.)

priv. Privately

prntng. Printing

rbbd. Rubbed (wear to covers; usually from handling)

rbckd. Rebacked

reprd. Repaired

reprnt. Reprint

rev. Revised

scttrd. Scattered

sgnd. Signed

shkn. Shaken (somewhat loose in covers)

sig. Signature

sl, slght. Slight

slipcase. Protective box

sm, smll. Small

sptd. Spotted

stmpd. Stamped

suppl. Supplementary

t.e.g. Top edge gilt (top edge has gold color applied)

tips. Corners

tndr. Tender (spine weakened)

trn. Torn

uncut. Untrimmed

unopened. Leaves not slit open in any manner

wrn. Worn

wrps. Wraps (covers made from paper rather than boards)

wtr stnd. Water stained

Approximate Book Sizes

American:

12mo: 7 3/4"

8vo: 9 3/4"

4to: 12"

folio: 13" or larger

British:

Cr 8vo: 7 1/2 X 5"

Ry 8vo: 10 X 6 1/2"

Imp 8vo: 11 X 7 1/2"

Demy (Dy) 4to: 11 1/2" X 8 3/4"

Foolscap (Fo Fol): 13 1/2" X 10"

Cr Fol: 15 X 10"

Ry Fol: 20 X 12 1/2"

Imp Fol: 22 X 15 1/2"

Terms Describing Condition

Fair: Less than good condition; with defects noted. Shows much wear but is not missing any pages.

G. Good: Shows wear but no major defects. May be lightly soiled, shelf worn.

VG. Very Good: Clean, tight copy. Defects noted. Still not pristine. (One might say, "cover has scratch, else very good.")

F. Fine: As new. No wear.

Dealer Index

AB
American Botanist
Keith Crotz
1103 W. Truitt Ave.
Chillicothe, IL 61523
(309) 274-5254

ABx
Anna Buxton
Redcroft 23 Murrayfield Road
Edinburgh EH12 6EP Scotland
031 337 1747

AT
Atelier Books
Ed Brodzinsky
P.O. Box 314
Schenevus, NY 12155
(607) 638-9962

BB
Brooks Books
Phil Nesty
P.O. Box 21473
Concord, CA 94521
(415) 672-4566

BF
Barbara Farnsworth
West Cornwall, CT 06796
(203) 672-6571

BLB
Beth L. Bibby Books
George Bibby & Nikki Wright
1225 Sardine Creek Road
Gold Hill, OR 97525
(503) 855-1621

CB
Carol Barnett
3562 NE Liberty
Portland, OR 97211
(503) 282-7036

DL
Daniel Lloyd
'Heather Lea', 4 Hillcrest Ave.
Chertsey, Surrey KT16 9RD, England
01 940 2512

GW
Gary Wayner
Rt. 3, Box 18
Ft. Payne, AL 35967-9501
(205) 845-5866

Hrt
Hortulus
139 Marlborough Place
Toronto, Ontario, Canada M5R 3J5
(416) 920-5057

JI
Jennifer Inderwick
P.O. Box 1526
Ganges, B.C. Canada V0S 1E0
(604) 537-4204

MD
Michael Dooling
72 North Street
P.O. Box 1047
Middlebury, CT 06762
(203) 758-8130

PBx
Pomona Book Exchange
Highway 52
Rockton, Ontario
Canada, L0R 1X0
(519) 621-8897

RS
Raymond M. Sutton, Jr.
P.O. Box 330
Williamsburg, KY 40769
(606) 549-3464

RW
Wilkerson Books
Robin Wilkerson
31 Old Winter St.
Lincoln, MA 01773
(617) 259-1110

WH
William Hecht
Box 67
Scottsdale, AZ 85252
(602) 948-2536

WW
Weldon & Wesley, Ltd.
Lytton Lodge
Codicote, Hitchin, Herts., SG4 8TE
England
Stevenage (0438) 820370